PRINCETON AERONAUTICAL
PAPERBACKS

1. LIQUID PROPELLANT ROCKETS
David Altman, James M. Carter, S. S. Penner, Martin Summerfield.
High Temperature Equilibrium, Expansion Processes, Combustion
of Liquid Propellants, The Liquid
Propellant Rocket Engine.
196 pages. $2.95

2. SOLID PROPELLANT ROCKETS
Clayton Huggett, C. E. Bartley and Mark M. Mills.
Combustion of Solid Propellants, Solid Propellant Rockets.
176 pages. $2.45

3. GASDYNAMIC DISCONTINUITIES
Wallace D. Hayes. 76 pages. $1.45

4. SMALL PERTURBATION THEORY
W. R. Sears. 72 pages. $1.45

5. HIGHER APPROXIMATIONS IN
AERODYNAMIC THEORY. M. J. Lighthill.
156 pages. $1.95

6. HIGH SPEED WING THEORY
Robert T. Jones and Doris Cohen.
248 pages. $2.95

PRINCETON UNIVERSITY PRESS · PRINCETON, N. J.

NUMBER 2

PRINCETON AERONAUTICAL
PAPERBACKS

COLEMAN duP. DONALDSON, GENERAL EDITOR

SOLID PROPELLANT ROCKETS

BY CLAYTON HUGGETT,
C. E. BARTLEY AND MARK M. MILLS

PRINCETON, NEW JERSEY
PRINCETON UNIVERSITY PRESS
1960

PRINTED IN THE UNITED STATES OF AMERICA

HIGH SPEED AERODYNAMICS

AND JET PROPULSION

———————•·•———————

BOARD OF EDITORS

   I. Thermodynamics and Physics of Matter. Editor: F. D. Rossini
  II. Combustion Processes. Editors: B. Lewis, R. N. Pease, H. S. Taylor
 III. Fundamentals of Gas Dynamics. Editor: H. W. Emmons
 IV. Theory of Laminar Flows. Editor: F. K. Moore
  V. Turbulent Flows and Heat Transfer. Editor: C. C. Lin
 VI. General Theory of High Speed Aerodynamics. Editor: W. R. Sears
VII. Aerodynamic Components of Aircraft at High Speeds. Editors:
A. F. Donovan, H. R. Lawrence
VIII. High Speed Problems of Aircraft and Experimental Methods.
Editors: A. F. Donovan, H. R. Lawrence, F. Goddard, R. R.
Gilruth
 IX. Physical Measurements in Gas Dynamics and Combustion.
Editors: R. W. Ladenburg, B. Lewis, R. N. Pease, H. S. Taylor
  X. Aerodynamics of Turbines and Compressors. Editor: W. R.
Hawthorne
 XI. Design and Performance of Gas Turbine Power Plants. Editors:
W. R. Hawthorne, W. T. Olson
XII. Jet Propulsion Engines. Editor: O. E. Lancaster

PRINCETON, NEW JERSEY
PRINCETON UNIVERSITY PRESS

PREFACE

The favorable response of many engineers and scientists throughout the world to those volumes of the Princeton Series on High Speed Aerodynamics and Jet Propulsion that have already been published has been most gratifying to those of us who have labored to accomplish its completion. As must happen in gathering together a large number of separate contributions from many authors, the general editor's task is brightened occasionally by the receipt of a particularly outstanding manuscript. The receipt of such a manuscript for inclusion in the Princeton Series was always an event which, while extremely gratifying to the editors in one respect, was nevertheless, in certain particular cases, a cause of some concern. In the case of some outstanding manuscripts, namely those which seemed to form a complete and self-sufficient entity within themselves, it seemed a shame to restrict their distribution by their inclusion in one of the large and hence expensive volumes of the Princeton Series.

In the last year or so, both Princeton University Press, as publishers of the Princeton Series, and I, as General Editor, have received many enquiries from persons engaged in research and from professors at some of our leading universities concerning the possibility of making available at paperback prices certain portions of the original series. Among those who actively campaigned for a wider distribution of certain portions of the Princeton Series, special mention should be made of Professor Irving Glassman of Princeton University, who made a number of helpful suggestions concerning those portions of the Series which might be of use to students were the material available at a lower price.

In answer to this demand for a wider distribution of certain portions of the Princeton Series, and because it was felt desirable to introduce the Series to a wider audience, the present Princeton Aeronautical Paperbacks series has been launched. This series will make available in small paperbacked volumes those portions of the larger Princeton Series which it is felt will be most useful to both students and research engineers. It should be pointed out that these paperbacks constitute but a very small part of the original series, the first seven published volumes of which have averaged more than 750 pages per volume.

For the sake of economy, these small books have been prepared by direct reproduction of the text from the original Princeton Series, and no attempt has been made to provide introductory material or to eliminate cross references to other portions of the original volumes. It is hoped that these editorial omissions will be more than offset by the utility and quality of the individual contributions themselves.

Coleman duP. Donaldson, General Editor

PUBLISHER'S NOTE: Other articles from later volumes of the clothbound series, *High Speed Aerodynamics and Jet Propulsion*, may be issued in similar paperback form upon completion of the original series in 1961.

CONTENTS

M. Combustion of Solid Propellants   3

    Clayton Huggett, Rohm and Haas Company, Philadelphia, Pennsylvania

    1. General Characteristics of Solid Propellants   3
    2. Thermal Decomposition of Propellant Components   7
    3. The Burning of Double-Base Propellants   21
    4. The Burning Rate of Propellants   30
    5. Theories of the Burning of Propellants   43
    6. The Mechanism of Burning of Composite Propellants   53
    7. The Ignition of Solid Propellants   57
    8. Cited References   61

H. Solid Propellant Rockets   64

    C. E. Bartley, Grand Central Rocket Company, Redlands, California
    Mark M. Mills, Radiation Laboratory, University of California, Livermore, California

    Chapter 1. General Features of Solid Propellant Rockets

    1. Introduction   64
    2. Outline of Construction and Operation   66
    3. Effect of Utilization on Rocket Design   73

    Chapter 2. Interior Ballistics Theory

    4. Scope of the Theory   77
    5. Combustion of Solid Propellants   77
    6. Stability of the Burning Surface   82
    7. Steady State Dynamics for End-Burning Grains   87
    8. Steady State Dynamics for Radial-Burning Grains   95
    9. Area Ratio Dependence. Erosive Instability   107
    10. Temperature Sensitivity, Transients, Thin Web Grains, Resonant Burning, Chuffing, and Gas Generation   117

    Chapter 3. Solid Propellants

    11. Composition and Preparation   123
    12. Propellant Properties   129

    Chapter 4. Design of Rocket Motors

    13. Discussion of Requirements   140
    14. Design of Propellant Grains   145
    15. Mechanical Design   154

    Chapter 5. Development Trends

    16. Trends in Solid Propellant Rocket Development   164
    17. Cited References   165

SECTION M

------◆------

COMBUSTION OF SOLID PROPELLANTS

CLAYTON HUGGETT

M,1. General Characteristics of Solid Propellants. Modern solid propellants used in jet propulsion devices may be divided into two general classes, double-base propellants and composite propellants, according to their composition and physical structure. Both types of propellants contain sufficient oxygen within their structure to effect their transformation into gaseous products. By the combustion or burning of a solid propellant we shall mean here the series of chemical processes through which the propellant, when subjected to a suitable ignition stimulus, is converted into gaseous products without the addition of oxygen from an external source. This combustion results in the evolution of a large amount of heat and, when it occurs in a confined space, in the development of a considerable pressure.

Double-base propellants have as their principal components nitrocellulose and an explosive plasticizer, usually nitroglycerin. Other materials may be added in smaller proportion to serve as stabilizers, nonexplosive plasticizers, coolants, lubricants, opacifiers, and burning rate modifiers, or otherwise to confer desirable properties on the product. Typical compositions together with some of their ballistic properties which will be discussed in later articles are given in Table M,1. The explosive plasticizer and minor constituents, with the exception of inorganic materials which may be present in small quantities, are physically compatible with nitrocellulose. The propellant is prepared in the form of a rigid plastic having a nearly homogeneous structure. Double-base rocket propellant compositions derive largely from those of the ballistites and cordites which have been used as gun propellants for many years. The principal innovations in adapting them for rocket applications have been in the techniques of fabricating the large grains needed for this purpose.[1]

Composite propellants are made by embedding a finely divided solid oxidizing agent in a plastic, resinous, or elastomeric matrix. The matrix

------

[1] A single piece of rocket propellant is called a grain although it may have a length of several feet, a web thickness of several inches, and weigh several hundred pounds. The web thickness is the minimum burning distance through the propellant between two opposite surfaces of the grain.

material usually provides the fuel for the combustion reaction although solid reducing agents are sometimes included in the compositions. Explosive binders requiring no additional source of oxygen have also been used. Composite-type propellants have been made in a great variety of compositions. Oxidizing agents which have been used extensively include ammonium nitrate, sodium nitrate, potassium nitrate, ammonium perchlorate, and potassium perchlorate. Asphalt, natural and synthetic rubbers, aldehyde-urea and phenolic resins, vinyl polymers, polyesters, and nitrocellulose are among the matrix materials which have been

Table M,1. Compositions and ballistic properties of typical rocket propellants.

Constituent	JPN Ballistite	HES 4016	Cordite SC	Russian cordite	Composite A	Composite B	Composite C
Nitrocellulose	51.5	54.0	50.0	56.5	21.0		
(Per cent nitrogen)	(13.25)	(13.25)	(12.20)	(12.20)	(12.60)		
Nitroglycerin	43.0	43.0	41.0	28.0	13.0		
Ethyl centralite	1.0	3.0	9.0	4.5	1.0		
Diethyl phthalate	3.25						
Dinitrotoluene					11.0		
Carbon black	(0.2)						
Potassium sulfate	1.25				9.0		
Potassium perchlorate					56.0		
Potassium nitrate							50.0
Ammonium perchlorate						75.0	
Ammonium picrate							41.0
Resin binder						25.0	9.0
Ballistic properties							
Heat of explosion, cal/g	1230	1260	965	880	1200		
Adiabatic flame temperature, °K	3028	3087	2492	2340		2200	
Burning rate; in./sec at 2000 lb/in.^{2}, 25°C	1.02	0.90	0.49	0.47	1.95	0.3	1.4
Pressure index at 2000 lb/in.^{2}	0.73	0.75	0.69	0.73	0.45	0.0	0.42

employed as binders. Ammonium picrate, carbon black, and aluminum powder have been used as fuel fillers. Minor constituents may be added to modify the properties of the binder or to catalyze the burning process. The common characteristic of these varied compositions is a markedly heterogeneous structure with adjacent regions of oxygen-rich and oxygen-deficient materials. In this respect they may be considered to be related to black powder although their physical form derives from modern plastics technology.

The laws of burning. Composite and double-base propellants show certain gross similarities in their burning behaviors. It is these properties

which they possess in common that largely determine their usefulness in rocket propulsion devices.

Solid propellants burn by parallel layers; that is, all burning surfaces regress in a direction normal to the original surfaces of the grain and the grain tends to retain its original configuration until the web has burned through. This generalization, known as Piobert's law, was first applied to granular black powder long before the discovery of smokeless propellants. It receives its most striking confirmation and has its greatest utility when applied to the large and geometrically intricate grains of modern rocket charges.

The linear rate of burning of a propellant depends on the pressure under which the burning takes place. The simplest form of this dependence states that the burning rate increases with some power of the pressure,

$$r = bp^n \tag{1-1}$$

where b and n are constants characteristic of the particular propellant composition under study. This generalization, known as Vieille's law, represents quite accurately the experimental results obtained with many simple double-base and some composite propellants. However, as indicated in Art. 4, the pressure dependence of the burning rate of many of the new propellant compositions is much more complex, particularly in the pressure region below 3000 lb/in.^{2}

A third generalization that will be found useful in considering the combustion of solid propellants may be stated as follows: The gaseous products from a propellant, burned under pressure in an atmosphere of its own combustion products, are in thermodynamic equilibrium at the temperature of the flame. Thus it is possible to calculate the composition and thermodynamic properties of the propellant gas from a knowledge of the composition of the propellant and the thermodynamic properties of its constituents, by methods which have been described in Sec. A. In any practical experiment the decrease in flame temperature due to heat loss to the walls of the combustion chamber, and to the expansion of the gas in the case of burning in a vented vessel such as a rocket motor, must be taken into account. More significant deviations from equilibrium will be observed when the burning takes place at low pressures (below perhaps 1000 lb/in.^{2}), when the burning takes place in a foreign atmosphere which may enter into or quench the reactions of the powder gas, and when the residence time of the reacting gas within the combustion chamber is too short for equilibrium conditions to be reached (Sec. A). Since these departures from equilibrium are due to the interruption of the reaction before it has had time to approach completion, they are indicative of conditions at some intermediate stage in the combustion process. The study of such nonequilibrium conditions, brought about by the

deliberate manipulation of experimental conditions, has provided much of the direct evidence concerning the nature of these intermediate steps.

Interior ballistic considerations. Before considering in detail the chemical processes which take place during the combustion of a solid propellant, it will be desirable to discuss very briefly the interior ballistics of solid propellant rockets. In this way we will be better able to evaluate from a practical standpoint the importance of the various factors which influence the burning process. A more detailed discussion of the interior ballistics of solid fuel rockets will be found elsewhere in this series (XII,H).

Most modern solid fuel rockets are designed around a neutral burning charge; that is, one whose burning surface area remains nearly constant until the charge is completely burned.[2] A simple example of such a charge consists of a centrally perforated, cylindrical grain with the end surfaces covered with an inhibiting material in order to confine the burning to the lateral surfaces. In this case the area of the burning surface within the central perforation increases at exactly the same rate as the area of the external surface decreases. The mass rate of burning of the propellant, or the mass rate at which the propellant gas is being generated within the chamber, will be given by

$$\dot{m}_b = S\rho r \tag{1-2}$$

where r is the linear burning rate, ρ is the density of the solid propellant and S is the area of the burning surface. We wish to consider the effect of variations in the chamber pressure p_c and the initial temperature of the powder T_0 on $\dot{m}_b$. Assume that the pressure dependence of the linear burning rate is given by Vieille's law (1-1).

$$\dot{m}_b = S\rho b p_c^n \tag{1-3}$$

The pressure index n is nearly independent of the propellant temperature but b increases as the initial temperature of the grain is raised.

At the same time that the grain is burning within the chamber, gas will be discharged from the nozzle of the rocket at a rate

$$\dot{m}_d = p_c C_d A_t \tag{1-4}$$

where A_t is the cross-sectional area at the throat of the rocket and C_d is called the discharge coefficient, a quantity whose value depends only on the composition and temperature of the propellant gas.

At the start of burning, the rate of gas generation will exceed the rate of gas discharge and the pressure in the chamber will rise. However, since n has a value of less than 1 for all practical propellants, the rate of gas

[2] Progressive charges in which the burning surface increases during burning, and regressive charges in which the burning surface decreases, may be used for special purposes. A great variety of grain designs have been used to obtain desired variations in the area of the burning surface with time during burning (XII,H).

discharge will increase more rapidly with pressure than the rate of gas generation, and eventually a chamber pressure will be reached where the two rates become equal and the chamber pressure assumes a constant value. At higher pressures, the mass rate of discharge would be greater than the rate of gas generation and the chamber pressure would fall to the equilibrium value. Thus a steady state pressure is established whose magnitude is, combining Eq. 1-3 and 1-4,

$$p_{eq} = \left(\frac{S\rho b}{A_t C_d}\right)^{\frac{1}{1-n}} \tag{1-5}$$

The pressure index may have a value in the neighborhood of 0.67 to 0.75 (Table M,1) so that the exponent $1/(1 - n)$ has a value of about 3 or 4. The equilibrium pressure is then seen to be an extremely sensitive function of S, the area of the burning surface and b, the coefficient of the burning law (it is assumed that ρ, A_t, and C_d are fixed for a given case). Since it is usually desirable to have the pressure in the rocket chamber show a minimum of variation under all circumstances, it is apparent that a suitable propellant must burn regularly in accordance with Piobert's law to give a constant burning area S and that b must show the smallest possible variation with propellant temperature. It is also apparent that the effects of variations in S and b will be minimized if n has a sufficiently small value. Thus the magnitude of the pressure index n is one of the important factors in determining the suitability of a propellant for rocket propulsion applications. The low value of n associated with many composite propellant compositions and the small variation of b with temperature are among the major advantages of this type of propellant.

M,2. Thermal Decomposition of Propellant Components. The combustion of a solid propellant takes place through a complex series of chemical processes which start within the interior of the grain and continue at an accelerated rate through the burning surface, finally reaching completion with the attainment of thermodynamic equilibrium in the gas phase at some distance from the surface of the propellant. The reaction zone possesses a physical discontinuity as well as sharp gradients in temperature and composition. The reactions are rapid, particularly at the high pressures which are of greatest interest. Consequently the precise experimental investigation of the burning process presents a difficult problem.

In order to gain some insight into the nature of the reactions taking place, it has been desirable to isolate experimentally certain probable reactions or groups of reactions. In particular the low temperature thermal decompositions of the oxidizing or explosive propellant ingredients have been studied extensively. While there is some question as to the applicability of the results of these *in vitro* experiments to the more complex conditions of combustion and to the multicomponent mixtures encountered in propellants, it appears that the reactions observed are similar to the initial steps of the combustion reaction. Moreover, the products of these low temperature decompositions must resemble the products from the initial stages of combustion, which in turn form the reactants for succeeding steps in the burning process.

Nitrocellulose. Nitrocellulose undergoes a slow spontaneous decomposition at temperatures ranging from those normally encountered in storage up to those where the sample ignites and rapid combustion follows. The reaction is catalyzed by traces of acids which may remain from the manufacturing process if the nitrocellulose is not sufficiently purified. Since some of the products of decomposition are themselves acidic in nature, the reaction may assume an autocatalytic character.

Will [1], in 1901, made careful studies of the effect of impurities on the rate of the decomposition reaction. He succeeded in showing that by careful purification the rate of the reaction could be reduced to a minimum value which could not be lowered further by additional purification. This characteristic rate increased with temperature and, for a given temperature, was greater the higher the degree of nitration of the nitrocellulose.

Robertson and Napper [2] studied the low temperature reaction in an apparatus wherein the volatile products of decomposition were removed continuously from the sample by a stream of carbon dioxide. They found that nearly half of the nitrogen liberated during this slow decomposition was in the form of nitrogen dioxide. If the decomposition products were allowed to remain in contact with the nitrocellulose, the NO_2 disappeared and the rate of total nitrogen evolution was accelerated.

Roginskii [3] has examined the data of Will and of Robertson and Napper and concludes that between 90 and 175°C the reaction follows the simple first order law (Sec. D).

$$-\frac{dc}{dt} = Ace^{-E_a/RT} \tag{2-1}$$

with an activation energy $E_a \cong 50$ kcal/mole and a frequency factor $A \cong 10^{20}$ sec^{-1}. Daniels [4] found an activation energy of 46.7 kcal/mole. Although a "normal" value of 10^{13} sec^{-1} for A is assumed, an examination of the data [81] indicates that the actual value must be in the neighborhood of 10^{19} sec^{-1}.

This slow decomposition is accompanied by the evolution of heat. When the rate of heat production at any point in the sample is greater than the rate of conduction of heat away from that point the temperature will rise, further accelerating the rate of decomposition and leading ultimately to a thermal explosion [5, p. 536]. Because of the low thermal conductivity of nitrocellulose, large samples stored at temperatures as low as 100°C may explode by this mechanism.

When small samples of nitrocellulose are heated above about 180°C, the sample ignites after a short induction period. Ignition is preceded by liquefaction and an acceleration of the decomposition process. A reactive gas containing nitrogen oxides, formaldehyde, carbon monoxide, and hydrogen is given off by the accelerated decomposition reaction. In the presence of an inert gas these products accumulate in the neighborhood of the nitrocellulose surface and a gas-phase ignition can occur, leading to a surface combustion of the sample. At very low pressures this reactive gas diffuses away so rapidly that ignition does not occur. Under these conditions the accelerated decomposition in the liquid phase can lead to a thermal explosion.

At very low pressures, under about 5 mm, small samples of nitrocellulose will burn with a feeble flame, depositing a powdery residue that has been termed the "white substance" (WS). The composition of WS has been studied by Wolfrom [6,7] and his associates. According to Wolfrom, a major part of this material is made up of partially denitrated and oxidized nitrocellulose residues having an average molecular weight of about 1500. This is accompanied by low molecular weight products such as formic and oxalic acids, furfural, glyoxal and formaldehyde, oxides of nitrogen and carbon, hydrogen, and water.

The decomposition becomes more complete at higher pressures. At 100 mm, about 40 per cent of the sample can be recovered as a complex mixture of liquid products called the "red substance" (RS) [6]. Glyoxal, formic acid, formaldehyde, and water are principal constituents of the RS, together with small amounts of glyceraldehyde, mesoxaldehyde, and other more complex products. The balance of the nitrocellulose is converted to gaseous products.

The following scheme has been proposed by Wolfrom, Dickey, and Prosser [7] for the initial reactions leading to the formation of RS.

$$
\begin{array}{ccl}
- - -\text{CH} & \quad & - - -\text{CH} \\
| & & | \\
\text{HCONO}_2 & & \text{HC}=\text{O} \\
| & \rightarrow & \| \\
\text{O}_2\text{NOCH} & & \text{O}=\text{CH} \quad + 3\text{NO}_2 \\
| & & | \\
\text{HCO}- - \;- - & & \text{HCO}- - \;- - \\
| & & | \\
\text{HCO}- & & \text{HCO}- \\
| & & | \\
\text{H}_2\text{CONO}_2 & & \text{H}_2\text{C}=\text{O}
\end{array}
\qquad (2\text{-}2)
$$

Cellulose nitrate

The first step is the breaking of the RO—NO₂ bonds, followed by rearrangement of the resulting free radical and scission of the 2—3 and 5—6 carbon—carbon bonds. By this scheme, glyoxal would originate from carbons 1 and 2 by hydrolysis of the acetal bonds. Formaldehyde would be formed from carbon 6 while the C—3,4,5 fragment would lead to the three-carbon aldehydes found among the products. Further reactions between NO_2 and these initial products would produce the more highly oxidized compounds found in the RS mixture.

Further increases in pressure cause a decrease in the amount of RS formed, but it is not until a pressure of several atmospheres is reached that all of the aldehydic products disappear. Fenimore [8] has formulated the final steps in the burning process as

$$HCO_2H, (HCO)_2, NO \rightarrow N_2, CO, CO_2, H_2O \qquad (2\text{-}3)$$

Nitroglycerin. The behavior of relatively large samples of nitroglycerin when heated at atmospheric pressure has been described by Snelling and Storm [9]. They observed that their samples began to discolor at 135°C. The color deepened and ebullition started at 145°C. This ebullition was due to the formation of gaseous decomposition products rather than to boiling, since on continued heating the temperature of the liquid rose to 218°C where explosion occurred. The nonexplosive decomposition is exothermic, as in the case of nitrocellulose, since the temperature of the sample rose above that of the surrounding air bath during the decomposition. By continued heating below the explosion temperature, a nonexplosive residue having a low nitrogen content was obtained.

Robertson [10] studied the low temperature reaction by a method similar to that which he had used to study nitrocellulose (Art. 2). He concluded that practically all of the nitrogen liberated during the decomposition appeared first in the form of nitrogen dioxide.

Roginskii and Sapozhnikov [11] carried out the decomposition in a closed system where the products of reaction remained in contact with the sample. At temperatures between 120 and 150°C an induction period during which a slow first order decomposition takes place was terminated by an abrupt transition to an explosive reaction. This was attributed to the accumulation of a critical concentration of some product of decomposition, probably NO_2, leading to an autocatalytic reaction. At temperatures between 155 and 190°C the reaction was first order, presumably because of the decreased solubility of the autocatalytic agent and the occurrence of a unimolecular decomposition in the gas phase. The activation energy of this first order decomposition was estimated to be 47.8 kcal/mole.

A very careful study of the autocatalytic reaction was made by Lukin [12]. He obtained explosions in sealed ampoules at temperatures as low as 70°C after long isothermal induction periods. Water and nitrogen dioxide, and oxalic, nitric, and sulfuric acids were among the active catalysts for the reaction. The activation energy for the autocatalytic process was found to be about 26 kcal/mole.

The explosive decomposition of nitroglycerin, as in the case of nitrocellulose, thus appears to be due to a combination of autocatalytic and thermal effects. A branching-chain type of explosion mechanism in the liquid phase appears unlikely in view of the failure of known chain-breakers such as benzophenone to inhibit the reaction [*3*] and the failure of ultraviolet light or electron bombardment to initiate long reaction chains [*13*]. However, Gray and Yoffe [*14*] present strong evidence for a chain mechanism for the vapor-phase inflammation of nitroglycerin and other alkyl nitrates from the accelerating effect of inert gas diluents on the reaction.

The decomposition of organic nitrates. While nitrocellulose and nitroglycerin are almost the only nitrate esters to be used extensively in solid propellants,[3] the low temperature decompositions of other members of the series have been studied in order to elucidate the mechanism of the process. In particular, the simple structures of the low molecular weight compounds make them more amenable to kinetic investigation.

A number of workers have suggested that the primary step in the reaction is the breaking of the RO—NO_2 bond with the resulting formation of a nitrogen dioxide molecule and an alkoxyl radical. The latter gives rise to aldehydic and hydroxylic products which then react further with NO_2 to give the observed final products, CO, CO_2, NO, N_2O, N_2, H_2, H_2O, etc. (Table M,2a).

The decomposition of methyl nitrate was studied by Appin, Todes, and Khariton [*17*]. They found the reaction to follow first order kinetics between 210 and 240°C and proposed the reaction scheme:

$$CH_3ONO_2 \rightarrow CH_3O\cdot + NO_2 \tag{2-4}$$
$$2CH_3O\cdot \rightarrow CH_3OH + CH_2O \tag{2-5}$$

At higher temperatures, explosion occurred following an induction period. The explosion was supposed to have resulted from a thermal process, but O. K. Rice [*18*] has questioned this interpretation of the experimental results. Gray and Yoffe [*14*] found that the addition of argon or nitrogen facilitated the explosion of methyl nitrate vapor, indicating a branching-chain mechanism for the process. Adams and Bawn [*15*] found the slow decomposition of ethyl nitrate to be similar to that of methyl nitrate and suggested an alternate scheme for the stabilization of the alkoxyl radical,

$$CH_3CH_2O\cdot + C_2H_5ONO_2 \rightarrow CH_3CH_2OH + C_2H_4ONO_2\cdot \tag{2-6}$$
$$C_2H_4ONO_2\cdot \rightarrow CH_3CHO + NO_2 \tag{2-7}$$

More recently, Phillips [*19*], and Pollard, Wyatt, and Marshall [*20*] have shown, through the retarding effect of NO_2 on the decomposition, that

[3] Diethylene glycol dinitrate was used extensively in Germany during World War II. This was due in part to a shortage of glycerin for the manufacture of nitroglycerin, but certain of the DEGN powders showed improved physical properties.

Table M,2a. Composition of the volatile products from the slow decomposition of nitrate esters.

Compound	Temp., °C	Gases, per cent								Liquid products	Reference
		NO_2	NO	N_2O	N_2	H_2	CO	CO_2			
Ethyl nitrate	185		60.0	3.2	1.2		8.9	23.4	CH_3CHO, H_2O	[15]
Nitroglycerin	165		37	5.4	} 11	3.9	22	30	H_2O	[12]
Nitrocellulose	160	10.1	37.2	5.4		3.9	13.8	19.7	CH_2O, H_2O	[5, p. 536]
Pentaerythritol tetranitrate	210	11.8	47.7	9.5	1.6	2.0	21.1	6.3	CH_2O, H_2O	[16]

the initial dissociation step (2-4) is reversible,

$$RCH_2O \cdot + NO_2 \rightarrow RCH_2ONO_2 \qquad (2\text{-}8)$$

Phillips has pointed out that the two-step mechanism (Eq. 2-6 and 2-7) is more probable than the simple disproportionation of alkoxyl radicals (Eq. 2-5) but Levy [22] has questioned the occurrence of this step because of the failure of diethyl peroxide to accelerate the decomposition of ethyl nitrate. The occurrence of free radicals during the thermal decomposition of nitrate esters has been discussed by F. O. Rice [32], but no experimental tests have been reported. The presence of free radicals has been demonstrated in the somewhat similar decomposition of ethyl nitrite [21]. In this case, the process

$$RCH_2O \cdot \rightarrow R \cdot + CH_2O \qquad (2\text{-}9)$$

apparently occurs when the reaction takes place at low pressures or in the presence of an inert gas diluent.

Most kinetic studies of nitrate ester decomposition have made use of manometric methods to follow the course of the reaction, assuming that the rate of pressure increase was proportional to the rate of disappearance of the nitrate. Analytical studies, if made, have been confined to the examination of the final products, and the nature of the intermediate steps has been inferred from the kinetic data and the products of decomposition. Recently, Levy [22] has used the infrared spectrophotometer to study the vapor-phase decomposition of ethyl nitrate. By this method he was able to follow directly the disappearance of the ethyl nitrate and the formation of some of the intermediate and final products. Ethyl nitrite was found to be formed as an intermediate in nearly quantitative yield. This decomposed subsequently to give acetaldehyde, ethyl alcohol, and nitric oxide, the products required by the older mechanisms (Eq. 2-4 to 2-8); so the failure of the prior experiments to reveal the presence of the nitrite as an intermediate is not surprising. Rate data obtained from the direct observation of the disappearance of the nitrate were in good agreement with those obtained previously by the manometric method.

The following series of equations provides a satisfactory explanation for the experimental observations:

$$C_2H_5ONO_2 \rightleftarrows C_2H_5O \cdot + NO_2 \qquad (2\text{-}10)$$
$$C_2H_5O \cdot \rightarrow CH_3 \cdot + CH_2O \qquad (2\text{-}11)$$
$$2CH_2O + 3NO_2 \rightarrow 3NO + 2H_2O + CO + CO_2 \qquad (2\text{-}12)$$
$$2CH_3 \cdot + 7NO_2 \rightarrow 7NO + 3H_2O + 2CO_2 \qquad (2\text{-}13)$$
$$C_2H_5O \cdot + NO \rightarrow C_2H_5ONO \qquad (2\text{-}14)$$

By carrying out the decomposition in the presence of added acetaldehyde or nitric oxide, the reverse step (Eq. 2-10) is suppressed and the rate of disappearance of ethyl nitrate follows simple first order kinetics.

If further work confirms the general occurrence of nitrites as intermediates in the thermal decomposition of organic nitrates, mechanisms concerned with the detailed chemistry of the decomposition process will have to be revised, although the over-all kinetic result as it affects the burning of propellants will probably be little changed.

Table M,2b. Thermal decomposition of organic nitrates.

Compound	Temp., °C	E_a, kcal/mole	$\log_{10} A$ †	Ref.	Remarks
Methyl nitrate	210–240	39.5	14.3	[17]	gas
Ethyl nitrate	180–215	39.9	15.8	[15]	gas
	161–181	41.2	16.85	[22]	gas
n-Propyl nitrate			14.7	[23]	
Ethylene glycol dinitrate	85–105	39	15.9	[23]	
Trimethylene glycol dinitrate	85–110	38.1	15.2	[23]	
Propylene glycol dinitrate	80–100	37.4	15.2	[23]	
Nitroglycerin	75–105	40.3	17.1	[23]	liquid
	90–125	42.6	18.0	[3]	liquid
	125–150	45	19.2	[3]	liquid
	150–190	50	23.5	[3]	liquid
Pentaerythritol tetranitrate	161–233	47	19.8	[16]	liquid
	171–238	39.5	16.1	[16]	5 per cent solution in dicyclohexyl phthalate
	100–120	50.9	20.6	[24]	solid
	120–150	54.0	23.3	[24]	liquid, 50 per cent in TNT
Mannitol hexanitrate			20.6	[23]	
Nitrocellulose	90–135	49	21	[3]	solid
	145–155	48	20	[3]	
	155–175	56	24	[3]	
	130–155	46.7	19[1]	[4]	
Nitroxy-nitro-compounds					
Trimethylolnitromethane trinitrate	75–95	36.4	15.3	[23]	
Dimethylolnitromethane dinitrate	134–208	38.9	16.5	[25]	

† $\log_{10} A$ estimated from published data [81].

The slow thermal decompositions of a large number of organic nitrates have been found to follow the first order kinetic law when precautions were taken to avoid the complicating effects of autocatalysis and self-heating. Values of the activation energy E_a and the frequency factor A in the unimolecular rate equation which have been reported by various investigators are collected in Table M,2b.

While there is some lack of agreement among different investigators studying the same compound, due probably to the experimental difficulties noted above, virtually all agree in assigning a high energy of activation and an unusually large frequency factor to these reactions. While the evidence is less conclusive, it also appears that A and E_a increase with the temperature and with increasing structural complexity of the molecule.

Large values of A and E_a are characteristic of most explosive substances (Table M,2b and M,2c). In particular, a large activation energy is necessary to confer adequate stability at low temperatures in combination with a high order of reactivity at elevated temperatures. The energy of activation for the decomposition of nitrate esters has been found to compare reasonably well with estimates of the strength of the $RO—NO_2$ bond.

The interpretation of the large value of A is more difficult. The possible occurrence of long, stable reaction chains (chain lengths of the order of 10^5 to 10^6 would be necessary to explain some of the reported results) is not supported by the experimental evidence, except possibly in the case of simple nitrate esters in the vapor phase. The more complex molecules show similar rates of decomposition in the liquid phase, in solution, and in the vapor phase (Table M,2b).

In the more complicated molecules, energy from a number of degrees of freedom of molecular motion may contribute to the activation of the $RO—NO_2$ bond, thus increasing the probability of bond rupture. Phillips [23] has suggested that the formation of the activated complex which precedes the breaking of the bond is accompanied by a large positive entropy of activation, due to a lessening of the repulsion between neighboring nitrate groups. While these considerations may account for a value of A well above the classical one, it seems probable that some of the very large values reported result from the complexity of the decomposition process and the difficulty in eliminating thermal and autocatalytic effects in the experimental measurements.

The decomposition of nitramines and nitro-compounds. While organic explosives other than nitrate esters have not been used extensively in propellants in the past, they appear to offer interesting possibilities for the development of new compositions. It has been shown that a variety of high explosives including TNT, PETN, and cyclonite will burn without detonation at rates comparable to those of propellant explosives.

The slow thermal decompositions of nitramines and nitro-compounds are similar in many respects to the decomposition of nitrate esters. Many of these reactions have been found to follow first order kinetics. The published kinetic data are collected in Table M,2c. The large activation energies and high frequency factors appear to be characteristic of all explosive compounds. Autocatalytic and thermal effects complicate the simple unimolecular picture as in the case of nitrate ester decompositions.

This may account for some of the discordantly large values of A and E_a reported for TNT and tetryl. Robertson [28] attributes the decreased rate of decomposition of cyclonite in solution to the quenching of a chain reaction which occurs in the pure liquid.

An examination of the products of decomposition reveals a more significant difference between the mechanism of decomposition of nitramines and of nitrate esters. While the primary step in the latter reaction

Table M,2c.   Thermal decomposition of nitramines and nitro-compounds.

Compound	Temp., °C	E_a, kcal/mole	$\text{Log}_{10} A$	Ref.	Remarks
Nitramines Ethylenedinitramine	100–120	37.3	12.2	[26]	solid
	120–135	41.5		[26]	solid
	135–145	51.5		[26]	solid
	184–254	30.5	12.8	[27, p. 677]	liquid
Cyclotrimethylenetrinitramine	213–299	47.5	18.5	[28]	liquid
	201–280	41	15.46	[28]	5 per cent solution in dicyclohexyl phthalate
	195–280	41.5	15.55	[28]	1 and 5 per cent solutions in TNT
Cyclotetramethylenetetranitramine	271–314	52.7	19.7	[28]	liquid
Nitro-nitramine 2,4,6-Trinitrophenylmethylnitramine	211–260	38.4	15.4	[27, p. 677]	liquid
	140–150	55.5	24.5	[3]	liquid
Nitro-compounds Trinitrotoluene	270–310	34.4	11.4	[27, p. 977]	liquid
		53	19	[3]	liquid
Picric acid		58.6	22.5	[3]	liquid
Nitromethane	380–430	53.6	14.6	[29]	gas

appears to involve the liberation of NO_2, followed by reduction to NO, most of the nitrogen from the nitramines appears first in the form of nitrous oxide, N_2O. The breaking of the RN—NO_2 bond to give nitrogen dioxide is not an important process in the decomposition of nitramines. This difference is illustrated by the difference in the composition of the gaseous products from decomposed nitramines and nitrate esters as shown in Tables M,2a and M,2d. Since the final step in the combustion

Table M,2d. Composition of the products from the slow decomposition of nitramines.

Compound	Temp., °C	Gases, moles/mole						Liquid products	Reference
		NO	N_2O	N_2	H_2	CO	CO_2		
Ethylenedinitramine	195	0.10	1.4	0.39	—	0.028	0.065	CH_3CHO, H_2O	[27, p. 677]
Cyclotrimethylenetrinitramine	225	0.54	0.98	1.16	0.09	0.40	0.48	CH_2O, H_2O	[28]
Cyclotrimethylenetrinitramine 5 per cent in TNT	220	0.70	0.63	1.65	0.20	0.57	0.83	CH_2O, H_2O	[28]
Cyclotetramethylenetetranitramine	280	0.95	1.51	1.16	—	0.57	0.64	CH_2O, H_2O	[28]

of an explosive appears to be the reduction, by carbon monoxide and hydrogen, of the oxides of nitrogen formed during the early stages of the reaction, we may expect certain differences between the combustion of nitramine and nitrate ester propellants because of the preponderance of either N_2O or NO in the gas-phase reaction.

A comparison of the data in Tables M,2b and M,2c indicates that, as a class, nitramines have greater stability at low temperatures than nitrate esters. From this it may be inferred that a higher surface temperature will be necessary for the combustion of a nitramine propellant. As shown in Art. 4, this suggests that the temperature dependence of the burning rate of a nitramine propellant should be low. No experimental tests of this sort have been reported.

The thermal decomposition of solid oxidizing agents. Of the various solid oxidizing agents which have been used extensively in rocket propellants, ammonium nitrate is the only one whose thermal decomposition has been studied in detail [30]. Robertson [16] found that the decomposition followed the unimolecular equation

$$k = 10^{13.8} e^{-40,500/RT} \text{ sec}^{-1}$$

over the temperature range 243 to 361°C. Nitrous oxide and water are almost the only products of decomposition of liquid ammonium nitrate at low temperatures. At higher temperatures NO_2, NO, and N_2 are found, probably as the result of a vapor-phase decomposition. By the use of N^{15} as a tracer, it has recently been shown that the nitrous oxide is formed from the reaction of an ammonium group and a nitrate group [31]. From the isotopic molecule $N^{15}H_4N^{14}O_3$, the isomer $N^{15}N^{14}O$ is formed exclusively. Friedman and Bigeleisen [31] suggest the dehydration mechanism,

$$NH_4^+ + NO_3^- \rightarrow H_2N{-}NO_2 + H_2O$$

$$H_2N{-}NO_2 \rightleftarrows HN{=}NO_2^- + H^+ \rightleftarrows HN{=}N\overset{\displaystyle O}{\underset{\displaystyle OH}{\diagup\diagdown}}$$

$$HN{=}N\overset{\displaystyle O}{\underset{\displaystyle OH}{\diagup\diagdown}} \xrightarrow[\substack{\text{or acid} \\ \text{catalyst}}]{\text{basic}} N_2O + H_2O$$

A variety of catalysts have been found to accelerate the decomposition of ammonium nitrate. Chromium compounds are particularly effective. This effect has been used to increase the burning rate of ammonium nitrate propellants, which otherwise are very slow burning [5, pp. 572–579].

The thermal decomposition of ammonium perchlorate has been studied by Dodé [32] and by Bircumshaw and Newman [92]. The reaction is complex; decomposition starts in the neighborhood of 200°C and is still slow enough to measure at 450°C. At temperatures below 300°C the major products of decomposition are indicated by the equation

$$4NH_4ClO_4 \rightarrow 2Cl_2 + 3O_2 + 8H_2O + 2N_2O$$

At temperatures above 350°C the following equation is more representative of the products:

$$2NH_4ClO_4 \rightarrow 4H_2O + Cl_2 + O_2 + 2NO$$

The interpretation of the experimental results is further complicated by the occurrence of an induction period whose duration is sensitive to the presence of impurities and by a change in crystal structure which takes place at 240°C. Manganese dioxide and ferric oxide catalyze the decomposition. Schultz and Dekker [93] have interpreted the experimental data of Bircumshaw and Newman in terms of a topochemical reaction starting at nuclei on the crystal surfaces and spreading hemispherically into the interior of the crystals.

Potassium perchlorate decomposes when heated to give potassium chloride and oxygen. The decomposition starts at about 500°C. Otto and Fry [33] report that the reaction is unimolecular over the temperature range from 536 to 617°C. From their results the rate equation

$$k = 10^{14.0}e^{-60,800/RT} \text{ sec}^{-1}$$

can be deduced. The large activation energy, combined with a normal frequency factor, is indicative of the stability of the material at low temperatures. A somewhat more complicated decomposition process is indicated by the more recent studies of Bircumshaw and Phillips [94]. The decomposition can be catalyzed by a variety of agents, ferric oxide and manganese dioxide being among the most effective.

Recently it has been shown that solid potassium perchlorate and carbon black react together at comparatively low temperatures, 320 to 385°C, at a rate much too fast for the normal dissociation of the perchlorate into potassium chloride and oxygen to be the rate-controlling step [34]. It is probable that similar processes take place in the burning of a composite-type propellant.

Sodium and potassium nitrate are also comparatively stable oxidizing agents. When heated they first lose a portion of their oxygen and are converted to the nitrites. This reaction is reported to start at 255°C in the case of $NaNO_3$ and at 286°C for KNO_3 [35]. At higher temperatures the nitrites decompose to form alkali oxides and peroxides. More oxygen and oxides of nitrogen are given off. The decomposition of KNO_3 is rapid at 1000°C but less rapid than the combustion of a KNO_3 propellant.

Gas-phase reactions. The varied observations on the slow decomposition of explosives suggest that the combustion of a propellant takes place through a surface zone decomposition leading to the formation of a combustible gas mixture. The combustion of this gas close to the propellant surface then provides much of the energy necessary to support the primary decomposition. The combustible mixture may contain reducing gases such as formaldehyde and other more complex organic fragments, carbon monoxide, and hydrogen; oxidizing agents such as nitrogen dioxide, nitric oxide, nitrous oxide, and oxygen; and the comparatively inert end products, nitrogen, water, and carbon dioxide. A knowledge of the kinetics of the various reactions that can occur in such mixtures would be of obvious value in attempting to arrive at an understanding of the complete combustion process. Unfortunately only a few of these reactions have been studied.

The reaction between nitrogen dioxide and formaldehyde has been studied by Pollard and Woodward [*36*, pp. 760, 767] and Pollard and Wyatt [*37*]. The reaction proceeds smoothly at temperatures between 118 and 160°C. The rate is proportional to $p_{CH_2O} \times p_{NO_2}$. The value of k is given by

$$k = 10^{7.1} e^{-15,100/RT} \text{ l/mole-sec}$$

Above 160°C the rate accelerates more rapidly, and at about 180°C the reaction becomes explosive. The slow reaction can be represented by the over-all equation

$$5CH_2O + 7NO_2 \rightarrow 3CO + 2CO_2 + 7NO + 5H_2O$$

The explosive reaction leads to the formation of large amounts of N_2 and CO_2 together with nitric oxide and a small amount of CO. The transition from slow reaction to explosion apparently occurs through a thermal mechanism.

The oxidation of CO by NO_2 is much slower than that of formaldehyde and thus is probably not important during the first stages of combustion where NO_2 is present [*38*].

In the later stages of the flame reaction NO, CO, and H_2 are the principal reactants. Nitric oxide is a relatively stable gas, decomposing only slowly into N_2 and O_2 at 1200°C [*39*]. It reacts slowly with carbon monoxide in a homogeneous third order process at 1380 to 1580°C [*40*]. At lower temperatures the reaction is heterogeneous and is catalyzed by water vapor [*41*]. The suggested mechanism is

$$CO + H_2O \rightarrow CO_2 + H_2$$
$$2NO + H_2 \rightarrow N_2O + H_2O$$
$$2N_2O \rightarrow 2N_2 + O_2$$

The reaction between NO and H_2 [*42*] takes place slowly at 900°C. It also follows a third order law, as do most reactions involving oxidation by NO.

M,3. The Burning of Double-Base Propellants. The many studies of the decomposition of pure explosives described in the preceding article suggest the types of chemical reactions that can be expected to occur in the combustion of a double-base propellant. This approach to the elucidation of the mechanism of the process has been stressed by Bawn [25] and by Adams [95, p. 277]. It must be kept in mind that these experiments deal with simple starting materials and, in general, with reaction in a single phase. Under these conditions we observe only a single reaction or closely related series of reactions. In the burning of a propellant, on the other hand, we encounter a series of reactions differing widely in chemical character and in the physical environment in which they occur. Any detailed

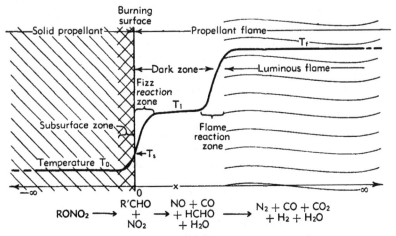

Fig. M,3a. Schematic representation of the combustion zone for a double-base propellant.

description of the burning process must take into account the physical and chemical interaction of these successive reaction steps.

Since the burning of a propellant takes place "by parallel layer," we may conveniently describe the burning as a one-dimensional process taking place along a spatial parameter normal to the burning surface. Because of the poor thermal conductivity of the material, the propellant at a short distance beneath the burning surface is largely unaffected by the combustion process. As the burning surface approaches, the propellant is heated by conduction and chemical reaction; the temperature rises, liquefaction may occur, and at the surface the propellant is completely decomposed to volatile fragments. These volatile products of decomposition react further in the gas phase and ultimately are converted to the end products of combustion at the high temperature of the propellant flame. This reaction sequence is illustrated schematically in Fig. M,3a.

It is convenient to choose a frame of reference so that the position of the burning surface remains fixed. As the propellant moves from left to right through the combustion zone, the products from each reaction step form the reactants for the following step until the end products, in thermodynamic equilibrium at the high temperature of the flame, are formed. Heat conduction in the opposite direction, from a region of higher to lower temperature, maintains the temperature of the reactants at a level necessary for reaction to occur. Thus a steady state is established, characterized by constant rates of mass flow and energy flow through the reaction zone and by stable concentration and temperature gradients. All derivatives with respect to time conveniently vanish. Such a one-dimensional model can be approximated experimentally in convenient fashion by a long cylindrical strand burning from one end. By coating the sides of the strand with some less combustible material, burning can be restricted to the end surface of the grain (Plate M,3). Such grains burn in a very regular and reproducible fashion and afford a convenient means of making a variety of observations on the burning process.

The subsurface zone. The propellant layers immediately beneath the burning surface are warmed by heat conducted from the relatively hot surface of the grain.[4] This rise in temperature starts the decomposition of the nitrocellulose and nitroglycerin a short distance below the surface. As has been shown (Art. 2), these decompositions are exothermic processes which cause further heating of the subsurface zone.

A second type of reaction which contributes to this subsurface heating is that which takes place between the nitrate ester and the stabilizer. These stabilizers are usually weakly basic aromatic compounds such as diphenylamine or ethyl centralite (*sym*-diethyldiphenylurea). It is their function to react with the oxides of nitrogen formed by the spontaneous decomposition of the nitrate esters during storage, thus preventing an accelerating autocatalytic reaction. At higher temperatures they can react directly with the nitrate esters in a highly exothermic reaction. The importance of this process is shown by experiments in which the temperature in the interior of a heated powder grain was measured by a thermocouple [43]. The temperature of the interior rose above that of the heating bath, and if the temperature of the bath was sufficiently high the sample ignited. The magnitude of this self-heating effect is greater for propellants containing relatively large amounts of stabilizer than for propellants with little or no stabilizer. Diphenylamine reacts more vigorously than centralite under these conditions and thus is a less satisfactory stabilizer

---

[4] We will neglect, for the present, the effect of radiation penetrating into the interior of the grain. This will be a good approximation in the case of a propellant containing an opacifying agent such as carbon black or in the case of a small end-burning strand burning at a low pressure where the volume of radiating flame is small and located at some distance from the burning surface. The effect of radiation in cases where it is of importance is discussed in Art. 4.

for high temperature storage. The importance of these reactions to the burning process is shown by the observation that small strands of propellants containing large amounts of stabilizers will burn smoothly in an inert atmosphere at atmospheric pressure and 35°C, whereas propellants containing only nitrocellulose and nitroglycerin will not burn at pressures below 200 lb/in.² at 35°C [44,84]. It appears that at low pressures only a small amount of energy returns to the burning surface from the flame zone. These exothermic reactions, taking place within the solid phase beneath the burning surface, must contribute a significant part of the energy necessary to decompose the surface layers of the propellant and transport the products to the flame reaction zone.

The heat balance in any plane of the combustion wave (Fig. M,3a) is given by

$$\frac{\partial}{\partial x}\left(k\,\frac{\partial T}{\partial x}\right) - \dot{m}\,\frac{\partial}{\partial x}\,(c_p T) + \dot{q}_x = 0 \qquad (3\text{-}1)$$

where x is the distance from the burning surface, k is the coefficient of thermal conductivity, $\dot{m}$ is the mass burning rate, c_p is the specific heat at constant pressure, and $\dot{q}_x$ is the rate of heat release due to chemical reaction in the x plane. The first term in the equation represents the rate of change in heat content per unit volume due to conduction, the second term is the rate of change of heat content due to mass flow, and the third term is the rate of heat release by chemical reaction. For purposes of simplification we assume that k and c_p are constant within the solid phase. With this approximation Eq. 3-1 becomes

$$k\,\frac{\partial^2 T}{\partial x^2} - \dot{m}c_p\,\frac{\partial T}{\partial x} + \dot{q}_x = 0 \qquad (3\text{-}2)$$

Unfortunately $\dot{q}_x$ contains the rates of the exothermic decomposition reactions. These depend on the temperature through an exponential factor of the Arrhenius type. The equation then could not be solved in exact form, even if the appropriate kinetic and thermodynamic terms for the evaluation of $\dot{q}_x$ were known. However, because of the large activation energies of these exothermic processes (Art. 2), $\dot{q}_x$ will be small outside of the high temperature region close to the burning surface. Qualitatively the effect of $\dot{q}_x$ will be to increase the steepness of the temperature gradient in the neighborhood of the burning surface.

If we make a further approximation, setting $\dot{q}_x = 0$, Eq. 3-2 can be integrated to give

$$T - T_0 = (T_s - T_0)e^{\dot{m}c_p x/k} \qquad (3\text{-}3)$$

where T_0 is the initial temperature of the propellant and T_s is the temperature at the burning surface. A typical double-base propellant may have the following properties:

$$\text{density} = 1.6 \text{ g/cm}^3$$
$$c_p = 0.35 \text{ cal/g deg}$$
$$k = 5 \times 10^{-4} \text{ cal/cm sec deg}$$

If we take $T_0 = 25°C$ and assume a value of 300°C for T, we can calculate T as a function of x for various values of the burning rate. The results of such calculations are shown in Fig. M,3b.

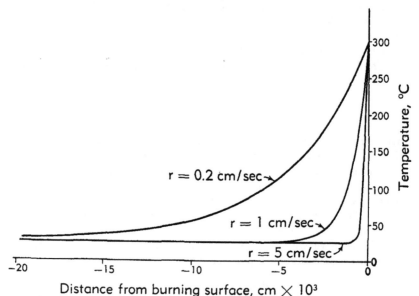

Fig. M,3b.   Calculated temperature profiles in the propellant
near the burning surface for various burning rates.

The shape of the temperature profile and the dependence of the thickness of the heated zone on the burning rate has been demonstrated experimentally by Klein, Mentser, von Elbe, and Lewis [45] through the use of very fine thermocouples. Some of their data are reproduced in Fig. M,3c.

It is apparent that only a very thin layer at the surface of the propellant grain takes part in the burning process. As the burning rate decreases, the thickness of this heated layer increases rapidly and the heat supplied by the exothermic reactions taking place therein makes up an increasingly large proportion of the total energy necessary to support combustion. In contrast to the transfer of energy from the flame region to the surface, the total amount of heat from this source may actually increase as the pressure decreases and the burning rate falls off. It is found experimentally that for most double-base propellants the pressure index of the burning rate becomes smaller as the burning rate decreases at low pressures. It is also generally observed that slow-burning propellants have smaller pressure indexes in the low pressure region than fast-burning ones. Since most jet propulsion devices operate at comparatively low pressures, below 2000 lb/in.², reactions in the subsurface zone may be of considerable practical importance and are deserving of more careful study.

The surface temperature. In the preceding paragraphs, reference was made to the temperature of the burning surface. It is apparent that this is a quantity which will be difficult to define and even more difficult to

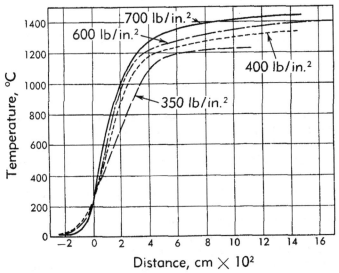

Fig. M,3c. Experimental temperature profiles in the neighborhood of the burning surface. Nitrocellulose (13.15 per cent N) plus 1 per cent ethyl centralite. The burning surface is assumed to be at $T = 250°C$ (from [45]).

measure experimentally. The temperature gradient is at its steepest in this region; thermal equilibrium between the various degrees of freedom of the reacting species may be upset by chemical excitation; and the location of the surface itself is rendered uncertain by liquefaction and foaming. Nevertheless, because of the utility of the concept of a surface temperature in discussing various characteristics of the combustion process, it is worthwhile to try to establish the approximate level of the temperature in the neighborhood of the burning surface. A number of experimental measurements of this quantity have been attempted.

From studies of the melting of small metal particles and the decomposition of alkaline earth carbonates when incorporated in propellants, Daniels, Wilfong, and Penner [4] estimated the surface temperature to be above 1000°C. This value appears to be too high since the surface of the burning grain is not luminous. The observed experimental results may be due to catalytic reactions at the surface of the solid particles, as has been observed with bare thermocouples in the propellant flame [45], or to variations in heat transfer to the solid particles which may project out from the propellant surface into the very steep temperature gradient of the flame.

The amount of heat in the surface layers of a suddenly extinguished propellant grain was measured calorimetrically by Aristova and Leipunskii [46]. If it is assumed that the temperature distribution within the burning surface is given by Eq. 3-3, the quantity of stored heat q is given by

$$q = \int_{-\infty}^{0} (T - T_0)\rho c_p dx = (T_s - T_0)\rho c_p \int_{-\infty}^{0} e^{\dot{m}c_p x/k} dx \qquad (3\text{-}4)$$

Then

$$T_s = T_0 + \frac{q\dot{m}}{\rho k} \qquad (3\text{-}5)$$

In this way a value of 330 ± 45°C was obtained for the surface temperature of a double-base propellant. A single-base (nitrocellulose) propellant gave 252 ± 48°C. The use of Eq. 3-3 assumes that no heat is released by chemical reaction within the solid phase. As this is not the case, the value of T_s given by Eq. 3-5 will be too small. Since these measurements were made at atmospheric pressure where the subsurface heating effect may be large, this may cause a significant error.

The direct measurement of the temperature profile by means of thermocouples, discussed in the previous article, also leads to a determination of T_s. Referring to Eq. 3-3 a plot of $\ln(T - T_0)$ vs. x should yield a straight line. This is found to be the case in the low temperature region of the temperature profile, but at higher temperatures a considerable curvature is observed [45]. This is due to the heat of reaction term which was neglected in obtaining Eq. 3-3. The point of departure from linearity represents the plane in the combustion zone where heat production becomes important. This is taken to be the burning surface. In this way a surface temperature in the neighborhood of 250°C was found for a single-base propellant. In this case also, a consideration of the heat of reaction term will lead to a somewhat higher value.

Finally it may be of interest to calculate an upper limit to the surface temperature from the measured rate of decomposition of nitrocellulose and nitroglycerin. Following Daniels, et al. [4], we estimate the number of nitrate groups in a typical double-base propellant to be $1.07 \times 10^{22}/cm^3$. For a burning rate of 1 cm/sec the surface reaction zone will have an effective thickness of perhaps 10^{-3} cm (Fig. M,3c). The rate of the uni-molecular decomposition reaction necessary to give a burning rate of

1 cm/sec will be

$$-\frac{dn}{dt} = kn = 1.07 \times 10^{22}/\text{cm}^2 \text{ sec} \qquad (3\text{-}6)$$

$$n = 1.07 \times 10^{19}/\text{cm}^2 \qquad (3\text{-}7)$$

$$k = 10^3/\text{sec} \qquad (3\text{-}8)$$

where n is the number of nitrate groups in the reaction zone and k is the specific rate constant. From Table M,2b

$$k = 10^{20}e^{-48,000/RT_s} \qquad (3\text{-}9)$$

and from Eq. 3-8 and 3-9 the value of the surface temperature[5] is:

$$T_s = 617°\text{K} = 344°\text{C}$$

Increasing or decreasing the thickness of the reaction zone by an order of magnitude leads to values of 310°C and 383°C respectively. Since the surface decomposition probably takes place, in part at least, by the faster autocatalytic mechanism, these values calculated for the simple unimolecular process must give an upper limit for the surface temperature.

While none of these results is completely satisfactory, it is probable that the temperature of the surface of a typical double-base propellant burning under moderate pressure is in the neighborhood of 300°C.

The flame reaction zone. When an end-burning strand of double-base propellant burns in an inert atmosphere at pressures below about 200 lb/in.², the flame is nonluminous [44,84]. As the pressure of inert gas is increased, a luminous flame region appears at some distance from the burning surface (Plate M,3). Further increases in pressure cause the luminous flame to approach the surface more closely, until at 1000 lb/in.² it is difficult to detect the dark zone between the propellant surface and the luminous flame. For a typical double-base composition the length of this dark zone between burning surface and luminous flame is given approximately by

$$l = \frac{10^7}{p^3}$$

where l is the dark zone length in inches and p is the pressure in lb/in.²

The heat produced by the nonluminous reaction, some 500 cal/g, is less than half of that produced by complete reaction through the luminous flame zone (Fig. M,3d). The products of reaction show a similar change in going from the nonluminous to the luminous stage (Fig. M,3e). Large

---

[5] Daniels calculates a much higher value for T_s by a similar process [4]. However, he uses a "normal" value of 3×10^{13} sec^{-1} for A, the temperature independent term in the rate equation rather than the experimental value used here. While Daniels does not calculate a value for A, an examination of his data [81], making reasonable assumptions concerning the nature of the products of decomposition, indicates that the value would not be far from 10^{19} sec^{-1}.

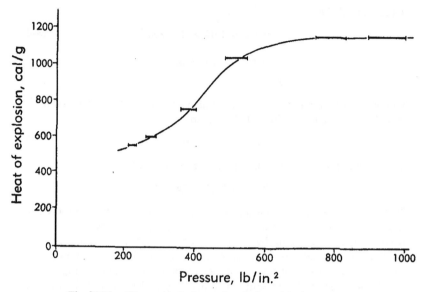

Fig. M,3d. Change in the heat of explosion with change in the
initial pressure of inert gas, propellant HES 4016 (from [44]).

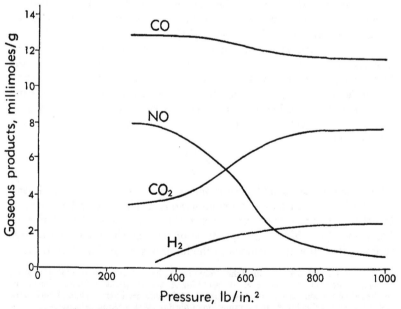

Fig. M,3e. Change in the products of combustion with change in the
initial pressure of inert gas, propellant HES 4016 (from [44]).

amounts of carbon monoxide and nitric oxide are formed in the non-luminous reaction. In the luminous flame zone the nitric oxide is consumed. The amount of carbon monoxide decreases, carbon dioxide increases, and hydrogen is formed by the water gas reaction.

The reacting flame can thus be divided into two regions with distinctly different properties and in which different chemical processes take place. The nonluminous region has been called the fizz zone since the product gases are projected outward in a direction normal to the plane of the burning surface with a considerable velocity, making a fizzing sound. The manner in which the products are projected outward indicates that the principal site of reaction must be very close to the burning surface. The luminous region is commonly called the flame zone to differentiate it from the fizz zone, but this usage is an unfortunate one since it may be misinterpreted to mean the gross volume of the propellant flame.

The temperature profile through the fizz zone of a nitrocellulose propellant has been measured by the thermocouple method described previously. Fig. M,3c shows experimental curves of temperature vs. distance for a number of pressures. As might be expected, the temperature gradient near the burning surface increases with pressure since the gas-phase reactions are accelerated. The fizz reaction zone is very narrow, extending only a few hundredths of a centimeter beyond the burning surface. This distance is much shorter than the length of the dark zone observed at comparable pressures. Beyond this narrow reaction zone the temperature levels off to a value of about 1400°C through the rest of the fizz zone. This latter figure is in satisfactory agreement with values found for double-base propellant by spectroscopic measurement of the intensity of nitric oxide absorption bands in the flame [85]. This latter method while less precise than the thermocouple method has the advantage that no foreign object is introduced into the flame. Calculations based on the measured heat of the fizz reaction and an analysis of the products lead to similar values for the temperature of the fizz zone.

From the results discussed in the previous paragraph it appears that the luminous flame reaction follows a period of comparative inactivity in the outer regions of the fizz zone. The reason for this delay in the onset of the final series of reactions is not known. We may compare this delay to the induction period frequently observed in gas-phase reactions. The depletion of some chain-breaking agent, the formation of a critical concentration of some autocatalytic intermediate, or the attainment of a critical temperature leading to a thermal explosion may bring about a sudden increase in reaction. No temperature profile measurements have been made in this region, but from the sharpness of the edge of the luminous flame we may infer that the gradient is rather steep. The temperature of the completely reacted flame has been measured by optical methods [87]. Good agreement with calculated values is found when heat loss in the experimental setup is taken into account. The final products of the reaction are in equilibrium at the temperature of the propellant flame.

Fig. M,3a summarizes the main features of this descriptive and qualitative picture of the burning process.

M,4. The Burning Rate of Propellants. The linear burning rate of a propellant and the manner in which the burning rate depends on the intrinsic properties of the propellant and on the conditions under which it is burned are factors of the greatest importance in determining the suitability of the propellant for jet propulsion applications. The elucidation of the factors which affect the burning rate is the primary objective and the principal justification for the study of the mechanism of the burning process. Conversely, a study of the effect of a given variable on the rate of burning can give important information on the combustion mechanism. In this article we shall follow this dual course: first, to describe the dependence of the burning rate on the intrinsic and extrinsic properties of the propellant and where possible to relate this dependence causally to the mechanism of burning; and second, to make use of the results of empirical burning rate studies to develop further the qualitative picture of the burning process presented in the previous article.

Experimental measurement of burning rate. Double-base propellants were used in guns for many years before they were widely used in rockets. It is not surprising therefore that the early measurements of burning rates were made under conditions simulating gun performance. The measurement of the variation of pressure with time in a gun chamber or, more conveniently, in a closed pressure chamber where the variation of chamber volume with shot travel does not occur, permits a determination of the linear burning rate. However, the calculation of the burning rate from such a pressure-time record is a laborious process which requires a number of assumptions and approximations. Indeed this calculation is seldom made; instead, a quantity called the "quickness," which involves both the linear burning rate and the geometry of the charge, is used to characterize and compare gun propellants. Moreover, the closed-chamber method of burning rate measurement is least satisfactory in the low pressure region of greatest interest in rocketry.

The extensive development of solid fuel rockets at the start of World War II led to more suitable methods for the study of rocket propellants. In the vented vessel technique the propellant charge is fired in a chamber fitted with a venturi nozzle, simulating a rocket motor. The variation of pressure with time is recorded. With a neutral-burning charge the chamber pressure remains fairly constant during burning. The linear burning rate at the average chamber pressure during burning is calculated directly from the web of the grain and the measured time of burning. Such a measurement is most useful in ballistic design work since the conditions under which the charge is burned can be made to approach closely those which prevail in a rocket in flight.[6] The method is less useful for the basic measurement of burning rates since the relatively large propellant grains required remove it from the realm of laboratory scale operations and the comparatively poor precision tends to obscure the small trends that are of importance in fundamental investigations.

The necessity for a true laboratory method of determining burning rates was recognized very early in the course of the study of the burning of rocket propellants. Such a method is provided by the strand burning technique [47].[7] In this method a long cylindrical propellant strand is ignited at one end in a pressure vessel which has been prepressurized to the desired level with an inert gas. The sides of the strand are coated with a suitable nonexplosive material to prevent the flame's spreading along the surface of the grain. The burning surface then remains planar and normal to the axis of the strand. Burning takes place at a nearly constant pressure and the burning rate is determined from the direct measurement of the time necessary to burn a measured length of strand. The method is rapid and precise and the small samples required are easily prepared in small scale equipment. The burning rate measured by this method is determined almost exclusively by the pressure and temperature at which the measurement is made and by the composition and physical structure of the propellant sample. It is thus well suited to the investigation of these variables. On the other hand, the application of data obtained by this method to the prediction of the performance of a charge burning in a rocket motor must be made with some care, since under the latter conditions the burning is influenced by the geometry of charge and chamber (see the latter part of Art. 4) as well as by the fundamental variables listed above.

Dependence of the burning rate on pressure. We have already noted that the linear burning rate increases strongly with an increase in the pressure under which the propellant burns, and that this pressure dependence at constant temperature is represented approximately by Vieille's law (Eq. 1-1),

$$r = bp^n \tag{4-1}$$

Here b, n, and a (Eq. 4-3 below) are constants characteristic of the propellant composition. They may vary with the temperature.

---

[6] However, these static tests in which the rocket test motor is fastened securely to a static test stand do not duplicate those forces on the propellant charge which are due to the motion of the rocket in flight. The setback force, due to acceleration of the rocket during burning, and centrifugal force, due to rotation in the case of a spin-stabilized rocket, will not affect the intrinsic burning rate of the propellant but they may have a serious effect on the over-all motor-charge performance due to displacement or breakup of the propellant charge.

[7] A number of ingenious methods which may be suitable for laboratory-scale investigations were also developed by British workers. These have not yet been described in the open literature.

Alternate forms of the burning law have been proposed. In the past the simple form

$$r = bp \qquad (4\text{-}2)$$

has been used in gun ballistics because it lends itself readily to ballistic calculations and because it represents the burning rate of simple single-base (nitrocellulose) gun propellants with some accuracy. However, it fails to describe the behavior of most rocket propellants adequately.

The linear form

$$r = a + bp \qquad (4\text{-}3)$$

has been advocated, particularly by Muraour and his associates [48]. This equation describes the behavior of many double-base propellants as well or better than the index law (Eq. 4-1). The two forms (Eq. 4-1 and Eq. 4-3) give similar results over the range of pressures usually of interest in rocketry, and both are used extensively in rocket ballistics calculations.

Careful measurements extending down to a pressure of a few atmospheres have shown that for many double-base propellants the burning rate can be represented very accurately over a wide range of pressures by the combined form

$$r = a + bp^n \qquad (4\text{-}4)$$

For most propellants a is small and n is not too different from unity so that Eq. 4-4 is approximated by either Eq. 4-1 or Eq. 4-3. For most heterogeneous propellants, for many homogeneous propellants containing various burning rate catalysts, and for most propellants at very low pressures, even Eq. 4-4 is inadequate to describe their complex behavior.

It is apparent that no single simple law can be put forth as *the* law of pressure dependence of the burning rate. However, the following generalizations can be made:

1. At high pressures (above a few hundred to one or two thousand lb/in.^{2}, depending somewhat on composition) the burning rate is a smooth function of the pressure and can be represented adequately by Eq. 4-1, 4-3, or more precisely by Eq. 4-4.
2. At intermediate pressures the pressure dependence may be complex and strongly dependent on composition. Regions of zero or even negative pressure dependence are sometimes found.
3. At very low pressures the burning rates of most propellants extrapolate to a finite rate at zero pressure although the propellant ceases to burn at pressures somewhere between a few pounds and a few hundred pounds per square inch.

In terms of the mechanism of the burning process, it appears that at high pressures most of the energy necessary to bring about the surface decomposition (which leads to a regression of the propellant surface and thus establishes the burning rate) comes from the high temperature flame region. The rate of energy transfer to the surface will be determined by the thermal conductivity of the flame and the proximity of the high temperature region to the burning surface. Since the rates of the flame reactions are strongly pressure dependent, this will lead to increased energy transfer and thus to a smoothly increasing burning rate as the pressure increases.

At intermediate pressures the flame zone is some distance removed from the burning surface or entirely absent. Much of the energy necessary to support combustion must come from the fizz zone. The reactions in this region are sluggish but susceptible to catalysis. They are strongly dependent on propellant composition, in contrast to the flame reactions which are virtually the same for all double-base propellants. In consequence, the pressure dependence of the burning rate is much less regular in this pressure region.

At very low pressures very little energy will return to the burning surface from the gas phase and the burning rate becomes nearly independent of pressure. Muraour attributes the pressure independent term in his burning rate law (Eq. 4-3) to thermal conductivity, independent of pressure. This can apply only to thermal conductivity in the condensed phase. We are then faced with the problem of accounting for the source of the energy necessary to maintain the high temperature of the burning surface, a source which must be *essentially independent of the pressure.* It is apparent that this energy necessary to support combustion at very low pressures (at atmospheric pressure or below) must come from the exothermic reactions taking place in the heated region below the burning surface.

It is necessary to point out that values of the limiting low pressure burning rate a, obtained by the application of Eq. 4-3 to burning rate data obtained in the high pressure regions, are not reliable because of the uncertainty of the pressure dependence at lower pressures. It is only by extending the burning rate measurements to very low pressures, preferably to atmospheric pressure or lower, that a value of a which may reasonably be interpreted in this manner is obtained.

Dependence of the burning rate on initial temperature. The change in burning rate with change in the initial temperature of the propellant is comparatively small for a chemical process, usually under 5 per cent per 10°C and frequently much less. Attempts to interpret this small temperature coefficient in terms of an activation energy for the rate-controlling process lead to implausibly low values for the energy of activation or improbably high values for the surface temperature. However, it is apparent from the nature of the burning process described in the preceding articles that such a simple interpretation cannot be expected to apply. The reaction does not take place in an isothermal region, but in a region of very steep temperature gradients where the precise application of the concept of an activation energy becomes difficult. The rate at which energy is supplied to the solid propellant may be considered to control the burning rate; the temperature in the reaction zone adjusts itself to maintain the steady state.

An increase in the initial temperature of the propellant will cause a similar rise, the exact value of which can be obtained from the specific heats of propellant and reaction products, in the final temperature of the flame. It might be assumed that the temperature profile in Fig. M,3a would be raised by an approximately constant amount at all points. However, the temperature of the burning surface cannot increase by this amount since, because of the increased burning rate, the surface layers have a shorter time to receive energy from the flame and subsurface reactions. The burning surface will then show a small increase in temperature but much less than the increase in T_0 or T_f.

For convenience in predicting the effect of an increase in T_0 we may relate the change in burning rate to the increase in energy transfer from the flame due to the increased differential between T_s and T_f or to the decreased amount of energy to be supplied to raise the powder from T_0 to T_s. Actually T_s assumes a value which represents a balance between these two factors, thus either method should lead to a similar result.

Corner [49] has developed the first method of approach. From the increase in T_f he calculates an increase in burning rate of 2.2 per cent per 10°C for cordite SC. The experimental value is between 3 and 4 per cent.

The second approach is implied in the use of the equation

$$r = \frac{c'p^n}{T' - T_0} \tag{4-5}$$

to represent the temperature dependence of the burning rate. Here T' is a constant, characteristic of the propellant composition and having the dimensions of temperature. This form for the temperature dependence, originally suggested by Crow and Grimshaw [50], has been applied by Avery and his associates [88,89] to a wide range of double-base propellants. Assuming for the moment that Eq. 4-5 represents the behavior of a propellant reasonably well, it can be seen that T' is the temperature (extrapolated) at which the burning rate would become infinite, i.e. the temperature at which burning would take place simultaneously throughout the propellant grain. This should be the surface temperature T_s. While it has not been possible to extend burning rate measurements to the neighborhood of T' because of the adiabatic heating of the sample, it is interesting to note that values of T' obtained by extrapolation from lower temperatures fall between 200 and 350°C, in reasonable agreement with estimated values of the surface temperature and the ignition temperature.

According to this argument a propellant with a high value of T_s should have a low temperature coefficient. It has been found experimentally that propellants which contain relatively stable coolants such as dinitrotoluene, and which might therefore be expected to have higher decomposition temperatures than a hot double-base composition, generally show a comparatively small dependence of burning rate on temperature. Many composite propellants, which because of the exceptional stability of their ingredients might be expected to have high surface temperatures, are also noteworthy for their small temperature coefficients.

The discussion in the preceding paragraphs has assumed that the burning rate is controlled by the final temperature of the flame T_f and that the effect of reactions in the solid phase can be neglected. As has been indicated, this is a good approximation at high pressures where the subsurface zone is narrow and the flame approaches close to the burning surface. It is not a satisfactory approximation at low pressures. Consequently it is not surprising that the temperature dependence of the burning rate at low and intermediate pressures is more complex than this simple picture would indicate. Propellants with zero or small negative temperature coefficients in this pressure region have been reported. The effects of compositional changes are often highly specific, but insufficient experimental data are available to permit useful generalizations. A careful study of the temperature dependence of the burning rate at low and intermediate pressures is very much needed.

Effect of propellant composition on the burning rate. The burning rate of a propellant at constant temperature and pressure is primarily a function of propellant composition. Changes in burning rate which accompany changes in composition may be divided into two classes: gross effects, which can be related to the change in the temperature of the propellant flame; and specific effects, which depend on a particular physicochemical action at some intermediate point in the burning process.

Muraour [51] was one of the first to point out the relationship between the burning rate and the flame temperature. He gives the equation

$$\log_{10} V = 1.36 + 0.27 \frac{T}{1000} \tag{4-6}$$

where V is the rate of decrease of the web (twice the burning rate as we have used the term) in mm/sec at a constant pressure of 1000 kg/cm² (14,223 lb/in.²) and T is the calculated uncooled temperature of the propellant flame. Using the linear form of the burning law

$$V = a + bp \tag{4-7}$$

where V is again the decrease in web in mm/sec and p is given in kg/cm², Muraour [52] finds that

$$\log_{10}(1000b) = 1.214 + 0.308\,\frac{T}{1000} \tag{4-8}$$

while a is nearly independent of composition and has a value of about 10 mm/sec. Eq. 4-6 and 4-8 are claimed to predict correctly the burning rate over a range of T_t from 1500 to 4000°C and p from 25 to 4500 kg/cm².

For most double-base compositions the flame temperature is approximately a linear function of the heat of combustion. It is therefore not surprising that a relationship similar to Eq. 4-6 should exist between the heat of explosion and the burning rate. Gibson [53] was the first to point out such a relationship. Muraour [54] gives the equation

$$\log_{10} V = 1.47 + 0.846\,\frac{Q}{1000} \tag{4-9}$$

where Q is the heat of explosion at constant volume in kilocalories per kilogram. Eq. 4-6 might be expected to show slightly better agreement with experiment than Eq. 4-9 since the rate of energy transfer to the burning surface depends more directly on T_t than on Q.

These empirical relationships were obtained from burning rate data taken at high pressures. From the mechanism of the burning process we may expect such a result to hold in the high pressure region where the rate of burning is largely controlled by energy transfer from the flame zone. We should hardly expect them to hold in the low and intermediate pressure regions where the hot flame zone is either absent or at some distance from the burning surface. Experimentally it is found that formulas similar to Eq. 4-6 or 4-9 hold reasonably well at pressures above some 2000 lb/in.² At lower pressures, propellants with similar heats of explosion may differ in burning rate by more than a factor of two. It is in these low pressure ranges that specific compositional effects become important.

Since at very low pressures the rate of burning is largely controlled by exothermic reactions taking place within the burning surface, we would expect compositions which favor such reactions to have relatively greater burning rates at these pressures. Propellants containing large amounts of stabilizers such as ethyl centralite or diphenylamine fall in this class and burn smoothly down to very low pressures. A related compound, p-phenylenediamine, is particularly active in this respect. A propellant to which 5 per cent of this material has been added burns faster at low pressures than the control sample without p-phenylenediamine despite the decreased heat of explosion. At higher pressures the burning rate falls below that of the control in accordance with Eq. 4-9.

More stable coolants such as dibutyl phthalate or triacetin do not enter into these low temperature reactions as readily; thus propellants containing these materials do not burn as readily at low pressures. A material which would react endothermally in the neighborhood of the burning surface would be expected to have an even greater retarding effect on the burning rate at low pressures. An experimental propellant containing paraformaldehyde is believed to illustrate this point [55,56,57]. It is probable that the paraformaldehyde dissociates to formaldehyde and then to carbon monoxide and hydrogen in the surface and fizz regions with the absorption of a large amount of energy. A propellant containing 5 per cent of paraformaldehyde failed to burn at low pressures but at higher pressures, as the luminous flame approached the burning surface, the burning rate increased rapidly with pressure and approached the value predicted from the heat of explosion.

Effect of radiation on propellant burning. Up to the present we have not considered the effect of radiation on the burning process. During the burning of a propellant, radiative energy transfer must occur between the hot flame and the propellant grain. According to the picture of the burning process which has been presented, this energy transfer must influence the burning rate. However, in contrast to the primary variables (pressure, temperature, and composition) which have been discussed previously, the radiation effect and other secondary effects to be discussed in this and the following articles depend on the geometry of the propellant charge and of the chamber in which it burns.

The effect of radiation on the burning process appears to be purely thermal in nature. No convincing evidence pointing to a specific photochemical reaction has been found.

Radiant energy absorbed by the propellant grain will raise the temperature of the propellant and thus increase its burning rate in a manner analogous to an increase in initial temperature. The effect of radiation on the burning rate can be represented by an equation similar to Eq. 4-5,

$$r = \frac{c'p^n}{T' - (T_0 + \Delta T_\lambda)} \tag{4-10}$$

where ΔT_λ is the increase in powder temperature due to absorption of radiation.

In case of a propellant having a large absorption coefficient at all wavelengths, the radiation will be absorbed in the surface layers and the effect will remain relatively constant (neglecting the effect of an increase in radiation path length) during the entire time of burning. If the propellant absorbs less strongly, radiation will penetrate into the interior of the grain. The propellant at the center of the web will be exposed to radiation for a longer time than that at the outside, so the burning rate will increase progressively during the entire time of burning. This effect has been discussed by Avery [58,59] and by Penner [59,60] who have made practical applications to the design of rocket charges. The increase in burning rate during burning, due to radiation, amounts to less than five per cent in typical rockets.

In extreme cases (transparent propellants with high flame temperatures) the intensity of radiation in the interior of the grain may become high enough to cause ignition at the surface of particles of absorbing impurities which are usually present [59]. This subsurface ignition causes fissuring and breakup of the propellant grain which may lead to dangerously high pressures in the rocket chamber. The addition of an opacifying agent such as carbon black to the propellant eliminates this source of erratic behavior.

The intensity of radiation at the burning surface is approximately

$$I = \sigma T_t^4 (1 - e^{-k\rho_g l}) \tag{4-11}$$

where σ is the Stefan-Boltzmann constant, T_t is the temperature of the propellant gas, k is the mass emissivity of the gas, ρ_g is its density, and l is the average radiation path length. This simple equation neglects the effect of emission and reflection from the comparatively cool propellant surface and from the chamber wall. It also assumes that the propellant gas is uniform in temperature and composition throughout the combustion chamber, while in fact the burning surface is separated from the hot flame by the comparatively cool fizz gas layer (Art. 3).

From Eq. 4-11 it is apparent that, because of the radiation path length term, the magnitude of the radiation effect will depend on the initial geometry of charge and chamber for a given propellant formulation. The radiation effect is strongly dependent on the flame temperature of the propellant, and to a lesser extent on the density of the flame gas which is proportional to the chamber pressure.

The simple molecules in the propellant flame are poor emitters so that the mass emissivity of the gas from a double-base propellant is low. The presence of a small amount of inorganic matter increases the mass emissivity greatly, probably due to the presence of incandescent solid particles in the flame. Estimates of the mass emissivity are not too satisfactory. Penner [60, p. 278] has given a value of $k = 25$ cm²/g for an unsalted propellant and $k = 40$ cm²/g for a propellant containing 1.5 per cent potassium sulfate. Experimental measurements by Craig [87] indicate even lower values.

An estimate of the importance of radiation in supplying the energy necessary to support combustion can be made. Consider a propellant with the following properties:

$$T_0 = 300°\text{K} \qquad c_p = 0.35 \text{ cal/g deg}$$
$$T' = 600°\text{K} \qquad \rho = 1.6 \text{ g/cm}^3$$
$$T_t = 3000°\text{K} \qquad k = 25 \text{ cm}^2/\text{g}$$

From Eq. 4-7 and 4-8 the burning rate of a propellant with a flame temperature of 3000°K at a pressure of 2000 lb/in.² will be approximately

1.5 cm/sec. The density of the flame gas at this pressure will be approximately 0.0125 g/cm³. We assume that all incident radiation is absorbed at the burning surface. Taking the mean radiation path length to be one centimeter, we get from Eq. 4-11

$$I = 29.6 \text{ cal/cm}^2 \text{ sec} \qquad (4\text{-}12)[8]$$

and

$$\Delta T_\lambda = \frac{I}{r\rho c_p} = 35° \qquad (4\text{-}13)$$

From Eq. 4-10

$$r = \frac{c'p^n}{600 - (300 + 35)} \qquad (4\text{-}14)$$

As a crude interpretation of these results, we can say that in this case radiation has increased the burning rate of the propellant by a little more than ten per cent or that radiation has supplied a little more than ten per cent of the energy which must be furnished to the burning surface to support combustion at the observed rate. While this figure will vary with the properties of the propellant (or with the nature of the assumptions just made), we note that in the most favorable case (large values of k, ρ_s, or l) the radiation intensity cannot exceed a limiting value set by the intensity of black body radiation at the temperature of the propellant flame. In the present case this is

$$I = 1.364 \times 10^{-12} \times 81 \times 10^{12} = 110.5 \text{ cal/cm}^2$$

and

$$\Delta T_\lambda = 131°$$

While radiation may play a significant role in the burning process, particularly in the case of hot propellants, it appears that it cannot play a dominant role in determining the burning rate.

Chuffing. It is observed that, when the pressure in a rocket motor decreases below a certain critical value, the chamber pressure may fall suddenly to atmospheric and the charge apparently ceases to burn. At times, however, the charge reignites after a delay lasting from a fraction of a second to many seconds and a new period of normal burning follows. This cycle may be repeated a number of times, leading to a series of explosions or "chuffs." The mechanism of the burning process which has been described in previous articles provides a qualitative explanation for this phenomenon.

The final stages of the gas-phase reaction have been shown to be sluggish at low pressures and to take place at some distance from the burning surface (Art. 3). If this reaction ceases within the rocket motor, as a result of a decrease in pressure due to regressive burning or to some

---

[8] T_f is the calculated flame temperature. In a typical rocket the actual temperature will be lower by perhaps 10 per cent. This will cause the radiation intensity to be lower than the values calculated here by a factor of $0.9^4 = 0.656$.

random disturbance within the motor, both the flame temperature and the number of moles of gaseous products will decrease suddenly. This will cause a sudden drop in the chamber pressure as well as a decrease in energy transfer to the burning surface. The temperature of the burning surface will fall and burning will cease.

However, the temperature of the surface layers of the propellant will still be relatively high; moreover, the surface will receive additional energy from the hot motor walls. Exothermic reactions, taking place within the surface layers, will lead to the formation of a heated zone penetrating a significant distance into the interior of the propellant grain. The exothermic decomposition of the surface layers leads to the formation of a combustible gas mixture which, when a critical concentration is reached, ignites spontaneously. The heated surface layers of propellant burn rapidly, maintaining the chamber pressure at a level high enough for the flame reaction to take place. As soon as this heated layer has been burned away, however, the burning rate decreases, the chamber pressure falls, and the cycle repeats itself. While most double-base propellants will cease to burn in a rocket when the chamber pressure falls below a critical value, the ability to reignite, thus producing a chuff, is seen to be closely related to the ease with which the propellant undergoes an exothermic decomposition, i.e. to the chemical composition of the propellant.

Andreev [61] has observed a similar type of pulsating reaction in the combustion of nitroglycol and has advanced a similar explanation.

Erosive burning. When a propellant grain burns under conditions such that there is a rapid flow of propellant gas parallel to the burning surface, the burning rate is enhanced. This condition may occur within a perforation in the grain where the products of combustion flow outward into the larger volume of the combustion chamber, or in a rocket chamber where the propellant gases flow over the propellant charge toward the nozzle located at one end. This phenomenon has been called erosion because of its superficial resemblance to the mechanical wearing away of a solid body by a rapidly flowing fluid. It is possible that a portion of the material at the burning surface, which may have a fluid or foamlike structure, may be carried away in this fashion. However, the principal effect appears to be due to increased heat transfer to the propellant surface, brought about by turbulence in the gas stream near the surface caused by the gas flow. The effect of erosive burning on rocket motor performance is discussed in detail elsewhere in this series (XII,H).

The rate of burning of a propellant under erosive conditions can be represented by the equation [90]

$$r = r_0(1 + kv) \tag{4-15}$$

or perhaps better by

$$r = r_0(1 + k\rho_g v) \tag{4-16}$$

where r_0 is the burning rate at zero gas velocity, v is the gas velocity parallel to the burning surface, ρ_g is the gas density, and k is the erosion constant characteristic of the particular propellant composition. The erosion constant is greater for propellants having a low flame temperature than for those with a high flame temperature; experimental values range from 10×10^{-4} sec/ft for a relatively "cool" cordite to 3×10^{-4} sec/ft for a "hot" ballistite. The erosion constant appears to be relatively independent of pressure; the temperature dependence of k is not known.

The gas velocity parallel to the burning surface will vary along the length of the propellant grain, being greatest at the end closest to the nozzle, and in a typical rocket may reach values of 500 to 1500 ft/sec. This will lead to a variation of the burning rate along the charge, causing the grains to assume a tapered shape, and to a significant increase in the average burning rate, factors of great practical importance in rocket design.

Corner [49,62] has given a simple theory of propellant erosion in guns, based on a simple one-step model of the burning process. From hydrodynamic considerations it may be concluded that the propellant gases in the chamber will be very largely in a state of turbulence. Close to the burning surface, however, there will be a laminar layer having essentially viscous properties. Outside this region a transition layer occurs wherein turbulence becomes increasingly important as the distance from the surface increases. Estimates of the thickness of the reaction zone indicate that the greater part of the flame reaction takes place in this transition region. Here heat transfer, which largely determines the burning rate, takes place both by convection and by conduction. A high gas velocity parallel to the burning surface increases the turbulence in the reaction zone, thus increasing the rate of heat transfer and the rate of burning. A cool propellant, having a thicker reaction zone, will have a larger portion of reaction occurring in the transition zone and thus will show a greater erosion effect than a hot propellant having a thinner reaction zone. This is in accord with experimental observations.

No attempt has been made to apply this theory of erosion to the more detailed models of the reaction zone which have been used in studying the burning process at the relatively low pressures encountered in rocket operation. The erosive burning of composite propellants has been studied by Green [96].

Resonance burning. The term "resonance burning" has been applied to a type of unstable burning sometimes encountered in rocket motors. This condition is found most frequently with tubular charges although other grain configurations have been reported to show similar effects. Resonance burning causes sudden rises in the rocket chamber pressure; these pressure peaks may be accompanied by breakup of the propellant grains, which in extreme cases may cause motor blowups. The central perforations of partially burned grains recovered from shots exhibiting resonance burning are found to have a rippled surface resembling a standing wave pattern. This first led to the suggestion that the unstable burning was caused by some form of resonance effect.

The occurrence of resonance burning appears to be closely related to the location of the stagnation point, the point of zero gas velocity within the grain perforation where the combustion gases flow outward in opposite directions. Resonance appears to be observed only when the charge geometry is such that the stagnation point occurs within the perforation of the grain. Empirically it has been found that resonance burning can be prevented by drilling radial holes in the grain, by placing a small noncombustible rod in the perforation, or by the use of perforations of noncircular cross section.

A limited amount of study has been given to the mechanism of resonance burning. Since the rate of burning of most propellants is a strong function of pressure and propellant temperature, a small fluctuation in pressure in the gas stream near the burning surface will cause a corresponding fluctuation in the burning rate at the adjacent surface. This will have the effect of reflecting the disturbance in an amplified form. The question of the origin of the initial disturbance has not been considered; however, all observations agree that small fluctuations due to small variations in composition or physical characteristics occur in all types of flames. It is only necessary to provide a suitable mechanism for their amplification to explain the present phenomenon.

The problem of a resonance effect between oscillations in the gas stream and the burning rate of the propellant has been studied by Grad [63]. It was found that under certain conditions the amplitude of such a disturbance would increase with time, leading to the possibility of increased burning rates, large pressure oscillations within the perforation, and grain breakup. The modes most favorable to such a resonance effect were found to consist of waves spiraling around the inside of the perforation. Experimental evidence supporting the acoustical theory of resonance burning has been presented by Smith and Sprenger [64, pp. 893ff.] who found that large amplitude pressure oscillations occurred within the rocket chamber during resonance burning with frequencies corresponding to the fundamental tangential mode of acoustical oscillation or its lower overtones. Similar results have been reported recently by Cheng [97], and his conclusions have been further discussed by Green [98].

The use of a noncylindrical perforation or of holes drilled through the sides of the grain will tend to break up the wave patterns and thus inhibit resonance burning. The inhibiting effect of an axial rod is less easily explained on the basis of this model. If the gas velocity is high the disturbance may be swept out of the perforation before it can reach dangerous amplitudes; in the neighborhood of the stagnation point, however, the gas velocity will be low so that the buildup time may be relatively long. Finally it may be noted that resonance effects are most severe with fast-burning "hot" compositions and that the burning rates of such powders are most sensitive to variations in temperature and pressure. The simple mathematical theory thus affords an explanation for most of the empirical observations concerning resonance burning.

M,5. Theories of the Burning of Propellants. The preceding articles, largely descriptive in nature, have given a qualitative picture of the burning process from which useful generalizations may be drawn. It would be desirable, however, to have a theory which would relate the burning rate and the basic parameters of the process, composition, pressure, and initial temperature, in a more fundamental manner. A successful theory would provide a detailed model from which properties of the reaction zone inaccessible to direct experimental measurement might be predicted and would provide a rational explanation for such phenomena as chuffing, erosion, and resonance burning, which up to now have been treated in a very speculative fashion.

Double-base propellants have been used in guns for a much longer time than in rockets, so that the early attempts to formulate a satisfactory theory of propellant burning were based on observations of the performance of propellants under gun conditions. The high pressures encountered in guns and closed chambers have the effect of shortening the scale of time and distance in the reaction zone. When the difficulties encountered in making accurate experimental observations under these conditions are considered, it is not surprising that these early theories were based on very simple models of the burning process. More recent investigations carried out in the low pressure regions have revealed the complexity of the process and demonstrated the necessity for treating a much more detailed model.

Corner [49] has reviewed the older theories of propellant burning, and more recently Geckler [95, pp. 289ff.] has summarized modern developments with particular reference to their applicability to the burning of composite propellants. Corner divides the earlier proposals into surface- and gas-phase theories depending on whether attention is directed at the initial gasification of the propellant surface or at the subsequent reaction of the combustible gas mixture. A complete theory will have to consider both processes and the interaction between the two. Theories of this type have been developed only within the last few years.

Surface theories. Of the surface theories, only that of Muraour [65] and a more recent hypothesis suggested by Daniels [4] need be considered here.

Muraour has not given a complete quantitative theory, but rather has correlated the results of a great variety of experimental observations in terms of a qualitative picture of the burning process. The surface decomposition of the propellant to give a combustible gas mixture is considered to be the rate-determining step, and the effects of such variables as pressure, initial temperature, flame temperature, heat of explosion, and radiation are interpreted as they affect this initial decomposition. Energy transfer from the flame to the propellant surface occurs by means of a conductive process whose rate is proportional to the pressure and by radiation, independent of pressure. This leads to the burning rate law

$$r = a + bp$$

While the theory, lacking a quantitative form, does not provide an insight into the detailed mechanism of the process it has been valuable as a guide to experimentation and as a framework for the correlation of experimental results. Some of the empirical correlations obtained by Muraour have already been cited (Eq. 4-6, 4-8, and 4-9).

Daniels has called attention to the first order decomposition of nitrocellulose and suggests that this is the rate-determining step in the surface decomposition, thus determining the rate of the whole combustion process. The nitrocellulose decomposition is assumed to have the experimentally measured activation energy, 46,700 cal/mole, and the "normal" frequency factor, 3×10^{13} sec^{-1}. This leads to a burning rate equation at constant pressure of the form

$$r = \frac{nAe^{-E_a/RT_s}}{m} \tag{5-1}$$

where n is the number of nitrate ester groups per square centimeter of surface reaction zone and m is the number of nitrate ester groups per cubic centimeter.

The effect of varying the initial temperature of the propellant is assumed to be reflected directly in the temperature of the burning surface (after making a suitable allowance for the increase in heat capacity of the propellant with temperature). This permits a calculation of the surface temperature, making use of the energy of activation and the experimentally determined temperature coefficient of the burning rate. In this way a value of about 1000°C is found for the surface temperature. The experimental evidence presented to support this high value of the surface temperature has already been discussed (Art. 3). In order to bring the burning rate calculated by means of Eq. 5-1 into agreement with experimental values, it is necessary to assume that the effective surface reaction zone is approximately one hundred molecules (5×10^{-6} cm) thick.

The effect of pressure on the burning rate cannot be derived from this hypothesis. Qualitatively it is assumed that an increase in pressure will speed up the (bimolecular) gas-phase reaction, bringing the reaction zone closer to the burning surface and increasing the heat flow to the surface. The theory is thus incomplete, as any theory which confines itself solely to the condensed phase must be, in that it tells nothing about the most significant step in the process, the flow of energy to the burning surface.

As a picture of the surface decomposition process, the greatest weakness of this hypothesis lies in the very large value assumed for the surface temperature. This in turn is made necessary by the use of a normal value for the temperature independent term in the decomposition rate expression despite the fact that all experimental evidence indicates that the true value (without attempting to interpret its possible significance at this time) is some 10^6 times greater.

Gas-phase theories. The surface theories concentrate their attention on the decomposition of the solid propellant at the burning surface and consider the flame zone only as an energy reservoir, with the rate of energy transfer from flame zone to burning surface depending on pressure and flame temperature in a simple fashion. In contrast, Boys and Corner [49,66,67] have considered the propagation of a reaction zone through a homogeneous combustible gas mixture. Their theory is applicable to any flame reaction which can be represented by a single-reaction step of thermal origin. In applying the theory to solid propellants, the manner of interaction between the flame zone and the burning surface is not considered. It is assumed that the propellant surface will decompose to produce a combustible gas mixture at a rate just equal to that at which it is consumed by the flame reaction. The manner of arriving at this steady state is not covered by the theory nor is the stability of such a state demonstrated; instead, the experimental facts are assumed to prove that such steady states do exist.

As a model for the combustion process, Boys and Corner consider the propagation of a plane reaction zone through a homogeneous combustible gas. The gas is assumed to be capable of undergoing a single reaction process whose rate at any point is determined only by the temperature and concentration of reactants at that point. The effect of diffusion of reactants in the reaction zone was neglected at the start but this effect has been included in a more recent paper [67]. The pressure is assumed to be effectively constant throughout the reaction zone,[9] thus the case of a gaseous detonation is not covered by the theory.

Consider a plane reaction zone moving along an x axis perpendicular to the plane of reaction with the origin moving with the reaction zone. The reactants approach the reaction zone from the region of $-x$, pass through the reaction zone, and the products move on in the direction $x \to \infty$. The conditions at any point in the reaction zone are a function of the x coordinate alone.

The total energy flux per unit area across any plane parallel to the reaction surface must be a constant since in the steady state there can be

[9] This can be shown to be true for slow flames to a very high degree of accuracy. See for example appendix B of [70].

no accumulation of energy between successive planes. This energy flux will consist of the intrinsic energy of the gas transported by mass flow, the work done in increasing the volume of the gas at constant pressure, and the energy flux due to thermal conductivity across the plane. At large distances from the reaction zone the temperature will be constant and the transfer of heat by conduction will be zero. This leads to the energy flux equation

$$\dot{m}(e + pV) - k\frac{dT}{dx} = \dot{m}(e_t + pV_t) = \dot{m}(e_0 + pV_0) \qquad (5\text{-}2)$$

or

$$\frac{k}{\dot{m}}\frac{dT}{dx} = h - h_t = h - h_0 \qquad (5\text{-}3)$$

Here $\dot{m}$ is the mass rate of flow per unit area (the mass burning rate), k is the thermal conductivity coefficient, h is the total heat content per unit mass, and the subscripts $_0$ and $_t$ refer to the initial and final states respectively. The heat content at any time when a fraction ϵ of the reaction is completed is

$$h - h_t = (1 - \epsilon)q + (T - T_t)c_p \qquad (5\text{-}4)$$

where q is the total heat of reaction per unit mass and c_p is the specific heat capacity, both at constant pressure. Since most of the reaction takes place in a comparatively small region at the high temperature end of the reaction zone, it is assumed that c_p, and thus q, is independent of temperature. Combining Eq. 5-3 and 5-4 we have

$$\frac{k}{\dot{m}}\frac{dT}{dx} = (1 - \epsilon)q + (T - T_t)c_p \qquad (5\text{-}5)$$

A second relationship between ϵ and T can be obtained from the kinetic expression for the homogeneous reaction rate. This will have the form

$$\frac{d\epsilon}{dt} = f(\epsilon, V, T) \qquad (5\text{-}6)$$

and since the rate at any point can be expressed as a function of x,

$$\frac{d\epsilon}{dt} = \dot{m}V\frac{d\epsilon}{dx} = f(\epsilon, V, T) \qquad (5\text{-}7)$$

Since Eq. 5-5 and 5-7 contain x only as dx, it is possible to eliminate x between them and obtain

$$\frac{d\epsilon}{dT} = \frac{kf(\epsilon, V, T)}{\dot{m}^2 V[(1 - \epsilon)q + (T - T_t)c_p]} \qquad (5\text{-}8)$$

Finally, V can be eliminated by means of the equation of state

$$\frac{pV}{RT} = \frac{1 + n\epsilon}{\mathfrak{M}} \qquad (5\text{-}9)$$

where a molecule of reactant of molecular weight $\mathfrak{M}$ gives $n + 1$ product molecules. Then

$$\frac{d\epsilon}{dT} = \frac{kp\mathfrak{M}f(\epsilon, T)}{\dot{m}^2 RT(1 + n\epsilon)[(1 - \epsilon)q + (T - T_t)c_p]} \tag{5-10}$$

This is an equation in ϵ and T where the one variable parameter $\dot{m}$ must be so chosen that it satisfies the boundary conditions $\epsilon = 0$ at $T = T_0$ and $\epsilon = 1$ at $T = T_t$.

Boys and Corner consider three particular cases of the general equation wherein the reaction is represented by (i) a first order rate resulting from a unimolecular reaction, (ii) a second order rate from a bimolecular reaction, and (iii) a second order rate as the low pressure result of a unimolecular reaction. The corresponding rate expressions are:

(i) $$\frac{d\epsilon}{dt} = f(\epsilon, T) = A_1(1 - \epsilon)e^{-E_a/RT} \tag{5-11}$$

(ii) $$\frac{d\epsilon}{dt} = f(\epsilon, T) = \frac{A_2}{V}(1 - \epsilon)^2 e^{-E_a/RT} \tag{5-12}$$

(iii) $$\frac{d\epsilon}{dt} = f(\epsilon, T) = \frac{A_3}{V}(1 - \epsilon)(1 + n\epsilon)e^{-E_a/RT} \tag{5-13}$$

By substituting these expressions in Eq. 5-10 and making use of Eq. 5-9, the corresponding equations for $d\epsilon/dT$ are obtained.

These latter equations are solved by a process of successive approximation. In cases (i) and (iii) explicit expressions for the mass burning rate are obtained.

Case (i) $$\dot{m}^2 = \frac{kA_1 p\mathfrak{M}e^{-E_a/RT_t}}{c_p RT_t(1 + n)\left[\frac{(1 + n/2)}{(1 + n)}\left(\frac{E_a}{RT_t}\right)\left(\frac{q}{c_p T_t}\right) - 1\right]} \tag{5-14}$$

Case (iii) $$\dot{m}^2 = \frac{kA_3 p^2\mathfrak{M}^2 e^{-E_a/RT_t}}{c_p R^2 T_t^2(1 + n)\left[\frac{(1 + n/2)}{(1 + n)}\left(\frac{E_a}{RT_t}\right)\left(\frac{q}{c_p T_t}\right) - 1\right]} \tag{5-15}$$

For case (ii) the solution has the form

Case (ii) $$\dot{m}^2 = \frac{kA_2 p^2\mathfrak{M}^2\psi e^{-E_a R/T_t}}{c_p RT_t^2(1 + n)^2} \tag{5-16}$$

where ψ depends on the composition of the fuel. The value of ψ is determined by the auxiliary equation

$$g_1(\psi) - \frac{2n}{1 + n}g_2(\psi) + \frac{n}{1 + n}g_3(\psi) = \left(\frac{c_p T_t}{q}\right)\left(\frac{RT_t}{E_a}\right) \tag{5-17}$$

This equation is solved for ψ by trial and error for a particular case, making use of the $g(\psi)$ functions tabulated by Boys and Corner.

The value of ψ is independent of pressure so that for all three cases, if the reaction can be represented by a rate equation of order n, the burning rate is proportional to the pressure raised to the $n/2$ power. This is a general result obtained with all flame theories based on a purely thermal reaction mechanism. The burning rate should be a smoothly increasing function of pressure unless the mechanism of reaction changes as the pressure rises. The theory fails to predict the irregular dependence of the burning rate on pressure frequently observed at low pressures.

With the value of the mass burning rate determined, the integration of Eq. 5-10 leads to a relation between ϵ and T. This in turn can be used in Eq. 5-5 to find a relationship between T and x. Thus the theory provides an insight into the structure of the reaction zone. The reaction zone is shown to be very narrow, with a steep temperature gradient and with most of the reaction taking place near the upper temperature limit. For a typical propellant, the reaction is estimated to go from 10 to 80 per cent toward completion in a distance of $1.6/p$ cm where p is the pressure in lb/in.²

In a later paper, Corner [67] has taken into account the diffusion of reactants and products in the reaction zone. The general features of the theory with diffusion are very similar to the simple theory in most respects, but it is found that inclusion of the diffusion effect can reduce the calculated flame speed by more than half in a typical case. Using the theory with diffusion and plausible values for the various reaction constants, including an activation energy of 26.4 kcal/mole, Corner obtained excellent agreement between observed and calculated burning rates for a series of propellants with flame temperatures ranging from 1900 to 3050°K.

A number of authors have developed theories of flame propagation in gases similar to that of Boys and Corner, but only the latter have attempted to apply their methods to the combustion of a solid propellant. While none of these gas-phase theories can be expected to give a complete account of the burning of a solid, they are applicable to that part of the propellant burning process which takes place in the flame zone. Theories of this type have been reviewed by Markstein and Polanyi [68] and by Evans [69]. Recently, Hirschfelder and Curtiss [70; 5, pp. 121, 127, 135] and their associates have developed their theory along the lines followed by Boys and Corner, but in a more general form. They obtained solutions to the flame equations for a number of reactions by numerical integration.

The flame theory of Zeldovich [71] and Frank-Kamenetsky has been applied by Belayev [72] and Semenov [73] to the combustion of nitroglycol. The liquid is assumed to volatilize at the surface, thus providing a constant supply of fuel for the vapor-phase flame reaction. The surface temperature is the boiling point of the liquid. It is stated that the theory is not applicable to nonvolatile explosives but it would appear that there is little difference, insofar as the mathematical formulation is concerned, between a simple vaporization and a surface decomposition of the type assumed by Boys and Corner.

These vapor-phase theories must be regarded as incomplete when applied to the combustion of a solid propellant in much the same way that the surface theories discussed in the last article are incomplete. They consider only the later stages of the reaction process, passing over the initial stages of the reaction with a simplifying assumption which is not compatible with the complexity of the process. These deficiencies of the older theories, made increasingly evident by recent experimental work, have led to the development of what may be called combination theories wherein an attempt is made to consider the interaction of all of the successive stages of the reaction process.

Combination theories. Theories of the combination type have been put forward by Rice and Ginell [55] and by Crawford and Parr [74]. The two theories are essentially similar, differing only slightly in the model which is assumed for the reaction zone and in mathematical detail. Both are based on a three-stage reaction zone similar to that described in Art. 3 and shown in Fig. M,3a. It is assumed that the homogeneous reaction rates of the surface, fizz, and flame stages can be represented by single expressions of the Arrhenius type similar to Eq. 5-11, 5-12, and 5-13, although each of these stages may consist of a complex series of individual reaction steps.

The nearly complete separation in space of the three reaction stages, which appears to occur at least at low pressures, permits each stage to be treated separately before considering its interaction with the adjacent stage. The rate of the surface reaction, expressed as a function of a surface temperature T_s, is considered first. The effect of the fizz reaction, with a maximum temperature T_1, on T_s is considered next. Finally the effect of the flame reaction, leading to the final flame temperature T_f, on T_1 and thus on T_s and on the burning rate, is considered.

Both theories assume a first order rate law for the surface reaction. Rice considers a surface decomposition of the Daniels type (Art. 5). From experimental observations the limiting low temperature value of the surface temperature T_s is estimated to be about 700°K. Values for the other parameters which determine the burning rate are then found by fitting the theoretical equations to experimental burning rate data. The low value of T_s together with the observed temperature and pressure coefficients of the reaction leads to a value for the energy of activation of the surface decomposition of some 16,000 cal/mole in a typical case. The frequency factor A_s is found to have a "normal" value in the range 10^{11} to 10^{13} sec^{-1}. These low values of E_s and A_s may be compared with the high values indicated by thermal decomposition experiments (Art. 2).

Crawford has extended Rice's model to consider a surface reaction zone of finite thickness, extending from the burning surface back into the powder to where the temperature has fallen to such a low value that further reaction can be neglected. Since the propellant decomposes to give gaseous products, the reaction zone is considered to have a foamlike structure with the density decreasing as the burning surface is approached. This foam zone is treated as a thermal reaction zone of the Boys-Corner case (i) type (Eq. 5-10 and 5-11), neglecting diffusion in the condensed medium.

The fizz zone is represented by a second order thermal process. Crawford follows the method used by Boys and Corner to treat such a process, with appropriately altered boundary conditions to allow for the proximity of the foam and flame zones on either side. Rice uses a somewhat different mathematical method to arrive at a similar result.

Crawford has not extended his theory to include the effect of the luminous flame but suggests that this effect will become important at high pressures. The reaction could be treated as a thermal reaction of the Boys-Corner type but, according to Crawford, it would have to be of a higher order than the fizz reaction to prevent coalescing of the two reaction zones. Rice considers the flame reaction to be a branching-chain explosion. It acts as a heat source whose distance from the burning surface depends on the time necessary to reach a critical concentration of some active species in the flame. For the purpose of fitting the theory to the experimental data this distance is obtained from measurements of the dark zone length (Art. 3). The flame reaction begins to make a significant contribution to the burning rate at a relatively high pressure, in the neighborhood of 1000 lb/in.² for a typical composition.

Since the burning rate is determined by a series of reaction stages having different pressure dependences and making their greatest contribution to the burning rate in different pressure regions, it is possible to reproduce the complex pressure dependence observed in the low pressure region by means of these theories. Fig. M,5 shows a comparison of the experimental burning rate data and the theoretical curves obtained by the two methods for a typical double-base composition. Except in the very low pressure region (surface reaction controlling) the theories of Crawford and of Rice give similar results. In the very high pressure region, where the flame reaction plays a dominating role, these theories can be made to merge smoothly with the gas-phase theory of Boys and Corner.

All of the theories discussed here require the evaluation of a number of reaction parameters which are not readily susceptible to direct measurement. Values are selected to give the best agreement between calculated and experimental burning rates but, because of the number of parameters involved and the approximate nature of the model considered, the selec-

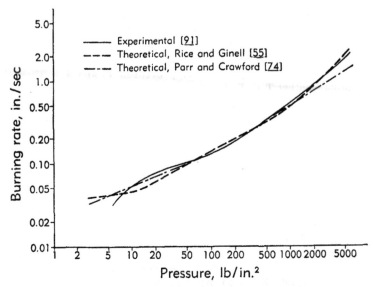

Fig. M,5. Comparison of experimental and theoretical
burning rate curves, propellant HES 4016, 25°C.

tion of a unique set of values is not possible. Table M,5 gives values of
some of the important reaction parameters used by various investigators
to obtain satisfactory agreement with experimental data.

Table M,5.   Values of reaction parameters.

	Boys and Corner [66]	Daniels, Wilfong, and Penner [4]	Rice and Ginell [55]	Crawford and Parr [74]
T_s A_s E_s	750°K	1273°K 3×10^{13} sec^{-1} 46,700 cal/mole	700°K 1.9×10^{11} sec^{-1} 16,000 cal/mole	700°K 20,000 cal/mole
T_1 A_1 E_1			1400°K $10^{-4}Z$ 5500 cal/mole	1100°K 30,000 cal/mole
T_f A_f E_f	3050°K $10^{-1}Z$ 26,400 cal/mole		3370°K	

As an example of the difficulties that are encountered in attempting
to obtain a consistent set of values in this fashion, we can consider the
surface decomposition reaction. If its rate can be represented by a first
order law of the Arrhenius type, three basic parameters, A_s, T_s, and E_s,

must be evaluated. If two of these can be established through independent measurements, the third can be obtained by comparison with the experimental burning rate. We can obtain a second relationship between T_s and E_s by means of the temperature coefficient of the burning rate (provided we can find a satisfactory relationship between T_s and T_0). Then the evaluation of either T_s or E_s, together with the observed burning rate and temperature coefficient of the burning rate, enables us to fix all three parameters. It should be pointed out, however, that measurements of the temperature coefficient are among the least satisfactory of the available data, from the standpoint of precision and accuracy, on which to base an evaluation of the burning parameters.

Daniels has taken his value of E_s from independent kinetic measurements of nitrocellulose decomposition. The value of T_s is then found by comparison with the observed temperature coefficient of burning although experimental evidence is also cited to support this value. A_s is then determined by the observed linear rate of burning. Rice has fixed his value of T_s on the basis of experimental evidence. The evaluation of E_s from the temperature coefficient and A_s from the linear burning rate then follows as before. Crawford has used experimental attempts to determine the parameters directly only as a guide to the appropriate range of values, and has varied both T_s and E_s to obtain the best possible fit to the experimental burning rate curves. In this way the three distinctly different sets of parameters in Table M,5 are obtained. Each of the three sets is self-consistent and capable of reproducing the observed burning rates with reasonable accuracy when applied to the appropriate form of the theory.

Much of the difficulty in evolving a completely satisfactory theory of the burning process must be attributed to our lack of knowledge of the individual chemical processes which take place. The theoretical treatment, while sometimes laborious in application, appears to be satisfactory for the simple models which have been proposed. Future advances will come through refinement of the model as new experimental work furnishes more information on the structure of the reaction zone and the nature of the chemical reactions comprising the successive reaction stages.

All of the theories discussed in detail in this article postulate a thermal model for the reaction zone, wherein the reaction is propagated by the conduction of thermal energy from the hot region of the flame to the cooler reactants entering the reaction zone. The importance of autocatalytic processes in the decomposition of propellants and explosives has been pointed out repeatedly, while chain reactions, propagated by free radicals or other energy-rich particles, are recognized to occur in most flames. Rice has suggested that the flame reaction is of the chain type, but this assumption is not essential to his theory.

Flame propagation theories in which the flame velocity is controlled by the diffusion of active chemical species rather than by the transfer of thermal energy have been developed by Lewis and von Elbe [75] and by Tanford and Pease [76].

The extension of the present theories to include reaction processes of these types appears to present no fundamental difficulties. An autocatalytic rate law could be substituted for the simple first order or second order laws assumed for the surface reaction zone of Crawford, the fizz zone of Rice or Crawford, or the flame zone of Boys and Corner, with only superficial changes in mathematical detail. The diffusion treatment of Boys and Corner, or of Rice and Ginell, can be adapted to the case of propagation by diffusion of active particles. Future developments will probably consider such reaction schemes, but here again substantial progress will be aided by further experimental elucidation of the role which autocatalytic and chain reactions play in the combustion process.

M,6. The Mechanism of Burning of Composite Propellants. Composite propellants have come into prominence only since the latter part of World War II. Consequently they have been studied less than the older double-base types. A recent review by Geckler [95, pp. 289ff.] provides an excellent summary of the present status of the problem.

The variety of compositional types falling within the broad definition of the composite class is so great that any general treatment of the burning process is probably out of the question. Nevertheless most composite propellants show similarities in physical structure and ballistic behavior which make it desirable to group them together as a single class. A comparatively low rate of dependence of burning rate on temperature and pressure is characteristic of many composite types.

The reaction zone in the combustion of a double-base propellant, as a consequence of the homogeneous structure of the solid, will be homogeneous in composition throughout any plane parallel to the burning surface. In contrast, the heterogeneous structure of the composite types leads to a similar heterogeneity at the burning surface and in the flame reaction zone. The thermal decomposition of oxidizer particles and matrix at the burning surface will give rise to alternate oxidizer-rich and fuel-rich gas streams. The manner of mixing of these oxidizer and fuel streams to form the flame reaction zone will play an important role in the burning process and may be largely responsible for the distinctive characteristics of the burning of composite propellants.

Crawford [91] and his associates have carried out experiments on the burning of a propellant composed of a stoichiometric mixture of potassium perchlorate and carbon black in a matrix of double-base propellant. Rice [57] has discussed the mechanism of burning of such materials in qualitative fashion. In this type of composite, the matrix in the absence of the filler is capable of burning in the manner described in Art. 3. The effect of the carbon-perchlorate filler is superimposed on the normal behavior of the matrix. The burning rates of propellants of this type increase rapidly with pressure in the region below 300 lb/in.², greatly exceeding the burning rate of the double-base matrix alone (Fig. M,6a). At higher pressures the burning rates of the composites appear to approach that of the matrix asymptotically, giving rise to the low dependence of burning rate on pressure, which is desirable in rocket applications.

At very low pressures the perchlorate appears to act as an inert diluent, cooling the burning surface by absorbing heat from the surrounding propellant, so that the burning rate falls below that of the matrix.

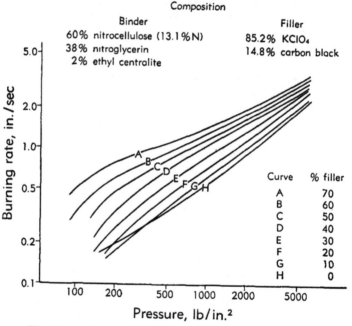

Fig. M,6a. Burning rates of composite propellants (from [91]).

Finely ground perchlorate is more effective than an equal weight of coarse material in lowering the low pressure burning rate, indicating that although the crystals do not reach thermal equilibrium in the burning surface the small crystals approach equilibrium more closely than the large ones.

As the pressure is increased, the perchlorate takes a more active part in the reaction. Photographs of burning grains show flashes of flame which appear to originate at perchlorate crystals. The perchlorate decomposes to give off streams of oxygen which can react with the carbon black and reducing gases in the flame. From the rate of decomposition of these crystals and published data on the rate of decomposition of potassium perchlorate (Art. 2), Rice has estimated the surface temperature of the crystals to be in the neighborhood of 1100°K. This is considerably higher than the probable surface temperature of the double-base binder. This high temperature may be caused by a reaction between perchlorate and the binder or binder decomposition products at the surface of the perchlorate particle. A more likely explanation is that the perchlorate particles extend out from the surface into the higher temperature region of the fizz flame. The diameters of the perchlorate crystals used are of the

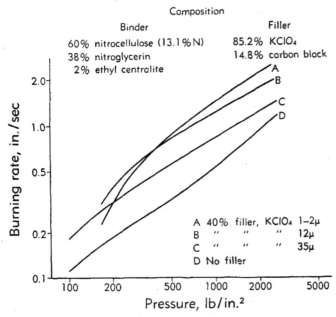

Fig. M,6b.   Effect of particle size of $KClO_4$ on the burning rate of composite propellants (from [91]).

same order of magnitude as the thickness of the fizz reaction zone. Photographs of burning grains suggest that only the larger crystals take part in the surface burning reaction at low pressures, while burning rate measurements show that coarse particles are more effective in accelerating the burning rate at lower pressures than an equivalent weight of fine particles of perchlorate (Fig. M,6b). If reactions at the surface of the crystal play a dominant role, the particle size effect should be reversed with the more finely divided material showing the greater effect at low pressures.

The decomposing perchlorate will give off streams of oxygen and particles of perchlorate may be ejected from the surface by gases generated at the particle-matrix interface, while the surrounding matrix will give off a combustible gas mixed with particles of carbon. The rate of reaction between these, and hence the distance from the hot flame zone to the propellant surface, will depend on the rate of mixing of the streams. Finely ground perchlorate will provide a great many tiny oxygen sources, thus promoting mixing by diffusion, decreasing the mixing distance, and increasing the burning rate by promoting energy transfer to the burning surface (Fig. M,6b). Rice has shown that this mixing distance will increase as the pressure increases so that the effect of this diffusion process is to decrease the fraction of heat returning to the burning surface and to resist an increase in burning rate as the pressure increases. This compensates to some extent for the strong pressure dependence of the matrix burning rate and leads to a more favorable pressure index for this type of propellant.

Propellants containing potassium perchlorate, ammonium perchlorate, or ammonium nitrate as the oxidizer have been prepared. When the compositions are adjusted to give propellants with similar flame temperatures, the order of the burning rates is $KClO_4 > NH_4ClO_4 > NH_4NO_3$. Surprisingly, the oxidizer with the greatest thermal stability gives the fastest burning rate while the least stable oxidizer gives the slowest burning rate.

The thermal decomposition of potassium perchlorate at the burning surface presumably leads to a gas mixture containing two parts of oxygen to one of potassium chloride (at flame temperature above the boiling point of KCl). Ammonium perchlorate decomposes to give a gas mixture containing large amounts of water vapor, chlorine, hydrogen chloride, nitrogen, and nitric oxide in addition to a small amount of oxygen (Art. 2). The principal products from the thermal decomposition of ammonium nitrate are nitrogen oxides, water vapor, and nitrogen. The relative burning rates of these composite types thus are found to parallel the oxygen content of the gas mixtures resulting from the thermal decomposition of the oxidizer rather than the ease of decomposition of the oxidizer.

The examination of the surface of partially burned grains of composites of the crystalline oxidizer-resinous binder type indicates that the solid- or liquid-phase reaction is confined to a very thin layer at the burning surface. Solid-phase reactions of the type which play such an important part in the burning of double-base propellants are not important here. The possible significance of reactions at the surface between matrix and oxidizer particles has not been explored.

This relative freedom from complicating solid-phase reactions, the possibility of studying the thermal decomposition of oxidizer and matrix independently, and the control which can be exercised over type, proportion, and particle size of oxidizer, makes this an attractive field for experimental and theoretical study. Further significant development of the theory of the burning of these relatively new propellant types can be expected.

M,7. The Ignition of Solid Propellants. The ignition of a solid propellant is a transient process leading to steady state burning. A discussion of the ignition process has been deferred until the nature of steady state burning could be examined.

In simplest terms, ignition consists in supplying energy from an external source to the propellant surface to produce a chemical and thermal state approximating that characteristic of steady burning. If the ignition system is well matched to the requirements of the propellant, energy will be supplied at a rate and in a quantity sufficient to lead to a close approximation of the final steady state. With the removal of the ignition stimulus, the normal combustion processes will take over the control of the reaction without discontinuity. If the energy supply from the igniter is not matched to the requirements of the propellant, various abnormalities such as overignition, igniter peaks, ignition delays, hangfires, and chuffing may occur in going from the transient to the steady state.

The ignition of gun or rocket propellant charges is usually accomplished by means of a small charge of black powder set off with an electric match. Less frequently, a metal oxidizer or other pyrotechnic mixture, finely divided propellant, or nitrocellulose is used as the main igniter charge.

In such a system, energy transfer to the surface of the propellant grain can occur principally through conductive and convective transfer from the hot igniter gas, through radiative transfer from the hot gas and incandescent solid particles, and through the impingement of hot solid particles from the igniter onto the propellant surface.

Ignition has been accomplished by each of these mechanisms under conditions such that the alternate methods of energy transfer could not play a significant part. The relative importance which the various mechanisms will assume in a particular case will depend on the composition of igniter and propellant and on the geometric parameters of the charge and igniter system. The use of black powder affords conditions favorable to all three mechanisms of energy transfer. The use of "gasless" igniters of the metal oxidizer type largely precludes the possibility of conductive and convective transfer from the gas. The effectiveness of this type of igniter must be due to the presence of hot solid particles coming in contact with the propellant surface and perhaps to radiation from the hot particles. When nitrocellulose or, more strikingly, an explosive gas mixture is used as the source of ignition energy, solid particles will be absent and the intensity of radiation from the gas will be low. Conductive transfer from the hot gas will be the principal mechanism of energy supply to the propellant surface.

Propellants can be ignited by radiation alone but the long ignition times observed are such as to suggest that radiation does not play a major role in the usual case. A consideration of the secondary role which radiation plays in steady state burning (Art. 4) supports this conclusion. The effect of radiation appears to be purely thermal; no photochemical effects have been observed. In the case of transparent propellants, radiation will penetrate into the interior of the grain. The distribution of absorbed energy will depend on the absorption coefficient of the propellant. Opaque propellants are found to be most easily ignited by radiation since most of the energy is absorbed in a thin layer at the surface, while propellants with some degree of transparency would be expected to burn more vigorously once ignition has been accomplished since the propellant will be heated to a considerable depth below the ignition surface. In the case of transparent propellants, radiation may even bring about a subsurface ignition due to the heating of radiation-absorbing particles of foreign matter unavoidably present in the propellant. Such subsurface ignitions occur after normal combustion has started and are caused by radiation from the propellant flame as well as from the igniter.

As is to be expected, propellants ignite more readily the higher their initial temperature. A rough correlation may be obtained between the minimum energy necessary to produce ignition in a given case and the energy necessary to heat the propellant from its initial temperature to some arbitrary ignition temperature.

The effect of pressure on the ignition process is more complex. High pressures will increase the emissivity of the igniter gas, thus increasing radiant energy transfer, but should have little effect on conductive transfer or on transfer by solid particles. The development of pressure will be accompanied by gas turbulence which may have an important effect on convective energy transfer, depending somewhat on the geometry of igniter and charge. Perhaps the most important effect of pressure in the ignition process lies in the transition to steady burning. The final gas-phase combustion reactions have been found to be unstable below a well-defined threshold pressure. If this pressure is not established in the combustion chamber during ignition, the transition to steady burning cannot take place smoothly and hangfire, chuffing, or an ignition failure will result.

Propellants with large heats of explosion are usually easier to ignite than cooler compositions. Propellants based on nitrate esters ignite more readily than those containing nitro- or nitramino-compounds. Double-base propellants ignite more readily than most composite types. A difference in ignition behavior is to be expected in the latter case because of the difference in burning mechanism.

The ignition of nitrocellulose has been studied by Rideal and Robertson [5, p. 536]. Two mechanisms can occur. If the gaseous products of thermal decomposition are allowed to accumulate in the neighborhood of the hot surface, a gas-phase explosion will occur leading to ignition.

This gas-phase reaction may be similar to the thermal explosion of formaldehyde and nitrogen dioxide studied by Pollard and Woodward [*36*, p. 767]. If the gaseous decomposition products are removed as they are formed, a gas-phase ignition cannot occur but on continued heating a thermal explosion will occur in the condensed phase.

The ignition of double-base propellants probably follows along similar lines. The importance of the retention of the ignition products in the neighborhood of the propellant surface has been illustrated by the work of Brian and McDowell [*77*] and others on the ignition of single grains of propellant by a stream of hot gases. The observed ignition times are longer than those observed for nitrocellulose under static conditions, and at times the grain vaporizes and the products are carried away by the gas stream without the appearance of a luminous flame. The presence of oxygen in the gas stream reduces the ignition time markedly. High speed photographs of igniting grains indicate that ignition starts in the gas stream close to the propellant surface. The occurrence of exothermic reactions in the condensed phase, which could lead to ignition by the thermal explosion mechanism, has been discussed previously (Art. 3).

Frazer and Hicks [*78*] have studied the properties of a thermal model for the ignition process. The propellant in the form of a semi-infinite slab with a plane surface at $x = 0$ receives heat at the surface from a high temperature gas representing the igniter flame. The propellant is also assumed to undergo a zero order exothermic reaction. Within the propellant grain the heat flow equation then has the form

$$c_p\rho \frac{\partial T}{\partial t} = k \frac{\partial^2 T}{\partial x^2} + qe^{-E_a/RT} \qquad x > 0; t > 0 \qquad (7\text{-}1)$$

in which c_p is the heat capacity, k is the coefficient of thermal conductivity, and q is the quantity of heat evolved per unit of reaction. At the surface being heated the boundary condition is

$$-k \frac{\partial T}{\partial x} = h(T_g - T) \qquad x = 0; t > 0 \qquad (7\text{-}2)$$

while at a sufficient distance from the surface the temperature gradient disappears and the other boundary condition becomes

$$-k \frac{\partial T}{\partial x} = 0 \qquad x \to \infty; t \geqq 0 \qquad (7\text{-}3)$$

The behavior of the igniter is idealized to the form of a thermal square wave; the temperature of the gas in contact with the propellant surface rises abruptly to a high value, $(T_g)_1$ at zero time, remains constant for a time t_0, and then falls to a low value $(T_g)_2$ for the duration of the experiment. This set of equations has been solved by numerical integration for a range of values of the parameters.

Ignition is arbitrarily defined to occur when the temperature of the propellant surface reaches a high value such that $T_i > 0.05E/R$, since at this point the exponential term in Eq. 7-1 is large and the temperature rise due to self-heating has become very rapid.

In this semi-infinite model, self-heating of the interior of the grain due to exothermic decomposition will always lead to ignition after a time t_{ia}, the adiabatic ignition time, without the addition of heat from an external source. The value of t_{ia} will depend strongly on the initial temperature T_0 of the propellant. In a real case, heat loss from the surface of a propellant grain of finite size will prevent an adiabatic ignition at initial temperatures below some critical value which depends on grain geometry as well as on the thermal parameters of the system. Frank-Kamenetsky [79] has derived the limiting conditions for adiabatic ignition for a number of simple geometries.

When heat is supplied to the propellant surface from an external source, the igniter gas at a temperature $(T_g)_1$, the surface temperature will rise and a thermal gradient will be propagated within the grain. If the heating is continued, ignition will occur after a time $(t_i)_{min}$, the minimum ignition time for the given rate of heat input. If heating is discontinued before ignition occurs $(t_0 < (t_i)_{min})$, heat will be lost from the surface to the cool gas at temperature $(T_g)_2$. However, if the propellant has reached a sufficiently high temperature before the hot igniter gas is removed, the temperature may continue to rise due to the exothermic decomposition reaction and the grain may ignite after a time $t_i > (t_i)_{min}$. As t_0 decreases below $(t_i)_{min}$ the ignition delay $(t_i - (t_i)_{min})$ increases rapidly and t_i approaches the adiabatic ignition time, corresponding to complete decay of the ignition stimulus.

The theory of Frazer and Hicks affords a satisfying picture of the thermal behavior of the propellant surface layers up to the point of ignition. The work of Rideal and Robertson, and others, seems to indicate clearly, however, that the site of the principal ignition process is in the layer of gaseous decomposition products adjacent to the propellant surface. If the definition of ignition used by Frazer and Hicks is extended by the assumption that the attainment of the ignition temperature is accompanied by the rapid evolution of an explosive gas mixture from the surface, this objection to the thermal theory is largely removed. The lack of suitable quantitative experimental data has prevented a rigorous test of the theory up to the present time.

The thermal theory of Frazer and Hicks with its emphasis on the solid-phase decomposition reactions is not applicable to most composite propellants where such reactions are largely absent. Jones [80] has studied the ignition of pyrotechnic mixtures similar in composition to certain composite propellants and has noted the absence of any heat of reaction term in the ignition energy equation. The ignition of a composite propellant by means of hot wires was studied by Altman and Grant [64, pp. 158ff.]. Thermal conduction was found to be the governing factor in this ignition process, heat generation by chemical reaction being relatively unimportant. The less frequent occurrence of chuffing or hangfires in the case of composite propellants also points to this difference between the mechanisms of ignition of composite and of double-base propellants.

With the heat of reaction term eliminated, Eq. 7-1 is reduced to the well-known unsteady state heat transfer equation

$$c_p\rho \frac{\partial T}{\partial t} = k \frac{\partial^2 T}{\partial x^2} \tag{7-4}$$

solutions of which have been tabulated for a variety of cases. The development of a theory of ignition of composite propellants along these lines with an accompanying experimental study appears to be highly desirable.

M,8. Cited References.

1. Will, W. *Angew. Chem. 14*, 743, 774 (1901).
2. Robertson, R., and Napper, S. S. *J. Chem. Soc. 91*, 764 (1907).
3. Roginskii, S. *Physik. Z. Sowjetunion 1*, 640 (1932).
4. Wilfong, R. E., Penner, S. S., and Daniels, F. *J. Phys. Chem. 54*, 863 (1950).
5. *Third Symposium on Combustion, Flame, and Explosion Phenomena.* Williams & Wilkins, 1949.
6. Wolfrom, M. L., Dickey, E. E., and Maher, G. G. The thermal decomposition of cellulose nitrate under reduced pressures. *Ohio State Univ. Research Foundation,* 1950.
7. Wolfrom, M. L., Dickey, E. E., and Prosser, H. C. A possible free radical mechanism for the initial reactions in the thermal decomposition, at reduced pressure, of cellulose nitrate (and other organic nitrates). *Ohio State Univ. Research Foundation,* 1948.
8. Fenimore, C. The final reactions of the burning of nitrocellulose. *Ballist. Research Lab. R-464,* 1944.
9. Snelling, W. O., and Storm, C. G. *U.S. Bur. Mines Tech. Paper 12*, 1912.
10. Robertson, R. *Proc. Chem. Soc. 25*, 179 (1909).
11. Roginskii, S. Z., and Sapozhnikov, L. M. *J. Phys. Chem. U.S.S.R. 2*, 80 (1931).
12. Lukin, A. Y. *J. Phys. Chem. U.S.S.R. 3*, 406 (1932).
13. Muraour, H. *Bull. soc. chim. France 53*, 612 (1933).
14. Gray, P., and Yoffe, A. D. *Proc. Roy. Soc. London A200*, 114 (1949).
15. Adams, G. K., and Bawn, C. E. H. *Trans. Faraday Soc. 45*, 494 (1949).
16. Robertson, A. J. B. *J. Soc. Chem. Ind. London 67*, 221 (1948).
17. Appin, A., Todes, O., and Khariton, Y. *J. Phys. Chem. U.S.S.R. 3*, 866 (1936).
18. Rice, O. K. *J. Chem. Phys. 8*, 727 (1940).
19. Phillips, L. *Nature 165*, 564 (1950).
20. Pollard, R. H., Wyatt, R. M. H., and Marshall, H. S. B. *Nature 165*, 564 (1950).
21. Rice, F. O., and Rodowskas, E. L. *J. Am. Chem. Soc. 57*, 350 (1935).
22. Levy, J. B. *J. Am. Chem. Soc. 76*, 3254, 3790 (1954).
23. Phillips, L. *Nature 160*, 753 (1947).
24. Brewer, S. D., and Henkin, H. The stability of PETN and pentolite. *Office Sci. Research and Develop. Rept. 1414,* 1943.
25. Bawn, C. E. H. The decomposition and burning of nitric esters. *British Ministry of Supply Rept. A.C. 10068/LFC. 58,* 1948.
26. Henkin, H. Stability of ethylenedinitramine (Haleite) and related nitramines. *Office Sci. Research and Develop. Rept. 1734,* 1943.
27. Robertson, A. J. B. *Trans. Faraday Soc. 44*, 677, 977 (1948).

28. Robertson, A. J. B.   *Trans. Faraday Soc. 45*, 85 (1949).
29. Cottrell, T. L., Graham, T. E., and Reid, T. J.   *Trans. Faraday Soc. 47*, 585 (1950).
30. Doescher, R. N.   *Calif. Inst. Technol. Jet Propul. Lab. Rept. PR 9-43*, 1950.
31. Friedman, L., and Bigeleisen, J.   *J. Chem. Phys. 18*, 1325 (1950).
32. Dodé, M.   *Bull. soc. chim. France 5 (5)*, 170 (1938).
33. Otto, C. E., and Fry, H. S.   *J. Am. Chem. Soc. 45*, 1134 (1923).
34. Patai, S., and Hoffmann, E.   *J. Am. Chem. Soc. 72*, 5098 (1950).
35. Lewchewski, K., and Degenhard, W.   *Ber. deut. chem. Ges. B72*, 1763 (1950).
36. Pollard, F. H., and Woodward, P.   *Trans. Faraday Soc. 45*, 760, 767 (1949).
37. Pollard, F. H., and Wyatt, R. M. H.   *Trans. Faraday Soc. 46*, 281 (1950).
38. Brown, F. B., and Crist, R. H.   *J. Chem. Phys. 9*, 840 (1941).
39. Wise, H., and Frech, M. F.   Rate of decomposition of nitric oxide at elevated temperatures. *Calif. Inst. Technol. Jet Propul. Lab. Rept. PR 9-46*, 1950.
40. Fenimore, C. P.   *J. Am. Chem. Soc. 69*, 3143 (1947).
41. Musgrave, F. F., and Hinshelwood, C. N.   *J. Chem. Soc. London*, 56 (1933).
42. Hinshelwood, C. N., and Green, T. E.   *J. Chem. Soc. London*, 730 (1926).
43. Crawford, B. L., Jr., and Isbin, H. S.   Determination of ignition temperatures of double-base powders. *Office Sci. Research and Develop. Rept. 1713*, 1943.
44. Crawford, B. L., Jr., Huggett, C., and McBrady, J. J.   *J. Phys. Chem. 54*, 854 (1950).
45. Klein, R., Mentser, M., von Elbe, G., and Lewis, B.   *J. Phys. Chem. 54*, 877 (1950).
46. Aristova, Z. I., and Leipunskii, O. I.   *Compt. rend. Acad. Sci. U.R.S.S. 54*, 503 (1946).
47. Crawford, B. L., Jr., Huggett, C., Daniels, F., and Wilfong, R. E.   *Anal. Chem. 19*, 630 (1947).
48. Muraour, H.   *Chimie & industrie 47*, 602 (1942).
49. Corner, J.   *Theory of the Interior Ballistics of Guns.* Wiley, 1950.
50. Crow, A. D., and Grimshaw, W. E.   *Trans. Roy. Soc. London A230*, 387 (1932).
51. Muraour, H.   *Chimie & industrie 50*, 105 (1943).
52. Muraour, H.   *Bull. soc. chim. France 9*, 511 (1942).
53. Gibson, R. E.   The rate of burning of double-base powders and the possible effects of change in nitroglycerin and total volatiles content on the burning of jet propulsion tube powder. *Office Sci. Research and Develop. Rept. 943*, 1942.
54. Muraour, H., and Aunis, G.   *Compt. rend. 229*, 173 (1949).
55. Rice, O. K., and Ginell, R.   *J. Phys. Chem. 54*, 885 (1950).
56. Rice, O. K.   The theory of the burning of double-base rocket powders. *Office Sci. Research and Develop. Rept. 5224*, 1945.
57. Rice, O. K.   The theory of the burning of rocket powders. *Office Sci. Research and Develop. Rept. 5574*, 1945.
58. Avery, W. H.   *J. Phys. Chem. 54*, 917 (1950).
59. Beek, J., Jr., Avery, W. H., Dreshner, M. J., McClure, F. T., and Penner, S. S.   Studies of radiation phenomena in rockets. *Office Sci. Research and Develop. Rept. 5817*, 1946.
60. Penner, S. S.   *J. Appl. Phys. 19*, 278, 392, 511 (1948).
61. Andreev, K. K.   *J. Phys. Chem. U.S.S.R. 20*, 365 (1946).
62. Corner, J.   *Trans. Faraday Soc. 43*, 635 (1947).
63. Gräd, H.   *Commun. on Pure and Appl. Math. 2*, 79 (1949).
64. *Fourth Symposium (International) on Combustion.* Williams & Wilkins, 1952.
65. Muraour, H.   *Mém. Artillerie Franç. 24*, 586 (1950).
66. Boys, S. F., and Corner, J.   *Proc. Roy. Soc. London A197*, 90 (1949).
67. Corner, J.   *Proc. Roy. Soc. London A198*, 388 (1949).
68. Markstein, G. H., and Polanyi, M.   Flame propagation—a critical review of existing theories. *Cornell Univ. Aeronaut. Lab. Bumblebee Series Rept. 61*, 1947.
69. Evans, M. W.   *Chem. Revs. 51*, 363 (1952).
70. Hirschfelder, J. O., and Curtiss, C. F.   *J. Chem. Phys. 17*, 1076 (1949).
71. Zeldovich, Y. B.   *J. Phys. Chem. U.S.S.R. 22*, 27 (1948). English transl. in *NACA Tech. Mem. 1282*, 1951.

72. Belayev, A. T.   *Acta Physicochim. U.R.S.S. 8*, 763 (1938).
73. Semenov, N. M.   *Prog. Phys. Sci. U.S.S.R. 24*, 433 (1940). English transl. in *NACA Tech. Mem. 1026*, 1942.
74. Parr, R. G., and Crawford, B. L., Jr.   *J. Phys. Chem. 54*, 929 (1950).
75. Lewis, B., and von Elbe, G.   *Combustion, Flames, and Explosions of Gases*. Academic Press, 1951.
76. Tanford, C., and Pease, R. N.   *J. Chem. Phys. 15*, 431, 433, 861 (1947).
77. Brian, R. C., and McDowell, C. A.   *Trans. Faraday Soc. 45*, 212 (1949).
78. Frazer, J. H., and Hicks, B. L.   *J. Phys. Chem. 54*, 872 (1950).
79. Frank-Kamenetsky, D. A.   *J. Phys. Chem. U.S.S.R. 13*, 738 (1939); *Acta Physicochim. U.R.S.S. 10*, 365 (1939).
80. Jones, E.   *Proc. Roy. Soc. London A198*, 523 (1949).
81. Daniels, F.   Studies of the mechanism of burning of double-base rocket propellants. *Office Sci. Research and Develop. Rept. 6559*, 1949.
82. Rice, F. O.   Presented at classified conference, 1950.
83. Roth, J.   Presented at classified conference, 1950.
84. Crawford, B. L., Jr., Huggett, C., and McBrady, J. J.   Observations on the burning of double-base powders. *Office Sci. Research and Develop. Rept. 3544*, 1944.
85. Crawford, B. L., Jr.   Spectrographic studies of the powder flame. *Univ. Minn. MR UMN/S2,3,4,5,7*, 1946.
86. Crawford, B. L., Jr.,   Measurement of flame temperature. *Univ. Minn. MR UMN/S9,10,12,13,14,15*, 1947.
87. Craig, R. S.   Flame temperature and radiation studies in rockets. *Office Sci. Research and Develop. Rept. 5832*, 1945.
88. Avery, W. H., and Hunt, R. E.   Effect of pressure and temperature on the rate of burning of double-base powders of different compositions. *Office Sci. Research and Develop. Rept. 1993*, 1943.
89. Avery, W. H., Hunt, R. E., and Sachs, L. D.   Revisions and corrections to Natl. Defense Research Comm. Formal Rept.a-225 (*O.S.R.D. Rept. 1993*, 1943). *Office Sci. Research and Develop. Rept. 4568*, 1944.
90. Thompson, R. J., and McClure, F. T.   Erosive burning of double-base powders. *Office Sci. Research and Develop. Rept. 5831*, 1945.
91. Crawford, B. L., Jr., et al.   Studies on propellants, 30, 35, 42. *Office Sci. Research and Develop. Rept. 6374*, 1945.
92. Bircumshaw, L. L., and Newman, B. H.   *Proc. Roy. Soc. London A227*, 115 (1954).
93. Schultz, R. D., and Dekker, A. O.   The kinetics of decomposition of ammonium perchlorate. *Paper presented at the American Chemical Society meeting, Atlantic City, Sept. 1952*. See also [95].
94. Bircumshaw, L. L., and Phillips, T. R.   *J. Chem. Soc.*, 703 (*1953*).
95. *Selected Combustion Problems: Combustion Colloquium*. AGARD, Palais de Chaillot, Paris, 1953.
96. Green, L., Jr.   *Jet Propulsion 24*, 9 (1954).
97. Cheng, S. I.   *Jet Propulsion 24*, 27, 102 (1954).
98. Green, L., Jr.   *Jet Propulsion 24*, 252 (1954).

SECTION H

SOLID PROPELLANT ROCKETS

C. E. BARTLEY
M. M. MILLS

CHAPTER 1. GENERAL FEATURES OF SOLID PROPELLANT ROCKETS

H,1. Introduction. Despite their fundamental simplicity, solid propellant rockets encompass a vast variety of detailed techniques. Just because of this simplicity, a realistic discussion of rockets requires a discussion of at least some of this technology, but the space available here is much too limited for a comprehensive treatment.

The choice has been made to emphasize the "power plant" aspects of the rocket. Propellant characteristics, theory of stability, internal ballistics, propellant grain, and combustion chamber design are discussed in some detail. On the other hand, vehicle problems, propulsive efficiency, missile accuracy, and related problems are mentioned only in a very qualitative way. Some discussion of these problems is required, because they often have a strong influence on the internal design of the rocket.

During World War II, an intensive rocket development effort was undertaken by most of the belligerents, chiefly aimed at the development of artillery rockets. Applications to long range missiles, guided missiles, and boosting, such as the assisted take-off of aircraft, have grown with the continuing improvements in the art of rocketry.

Artillery rockets, rocket-propelled missiles employed in much the same manner as gun artillery, have become important as an adjunct to guns in mobile warfare. Such missiles may range from 2 to 14 in. in diameter, from 20 to 120 in. in length, and from 3 to 1500 lb in weight. Since the rocket launcher weighs about the same as one round of rocket ammunition, artillery rocket weapons are much lighter and more mobile than corresponding guns. The rocket launcher does not have to confine the propelling gases, as the gun tube, nor does it have to absorb a large recoil; it serves merely as an aiming device. For example [1], the 75-mm gun fires a 13.5-lb missile and weighs nearly 2700 lb. An artillery rocket with a 14-lb payload would weigh about 24 lb total, and could probably be fired from a 26-lb launcher. Artillery rockets are usually inferior to guns in range, accuracy, and target penetration; but their superior mobility can compensate for this in many circumstances.

Rockets fired from aircraft, because the aircraft velocity may be added to the rocket velocity, are nearly as effective as a gun-fired shell of the corresponding caliber. A medium bomber can fire a salvo of 5-inch diameter rockets equivalent to a destroyer broadside. Rocket fire power can be made very large by utilizing multiple launchers or automatic launchers. Firing rates of 200 rounds of 5-inch missiles per minute are easily obtained.

Boosting applications are becoming increasingly important. The take-off of fast aircraft, otherwise requiring long runways, may be assisted by solid propellant rockets. Missiles frequently employ solid propellant booster rockets for launching, with sustaining thrust supplied by a variety of power plants such as the ramjet, the turbojet, and the liquid propellant rocket or solid propellant rocket.

Solid propellants obtain their name from the fact that they have mechanical properties resembling the mechanical properties of solid materials. Some propellant materials are actually very viscous fluids with peculiar thixotropic characteristics. A combination of properties is usually desired. A fairly rigid material is necessary so that the burning surface may be accurately defined and controlled, but a moderate degree of elasticity is required to permit the charge to accommodate itself to pressure and thermal stresses without cracking.

The chief advantage of solid propellant rockets, as compared to liquid propellant rockets, is their greater ease of utilization. After preparation, at the point of manufacture, they are easily transported and may usually be fired with simple electrical circuits. Elaborate field servicing equipment is usually not needed. For some applications, solid propellant rockets also can be superior to liquid propellant rockets in total impulse-to-total weight ratio.

The major advances in the development of solid propellant rockets during World War II were: the development of new propellants which were more suitable than gun propellants or black powder for rocket propulsion, the discovery and quantitative analysis of special characteristics of these propellants which strongly affect rocket design and performance, the invention of propellant materials with good mechanical properties capable of withstanding pressure and acceleration forces without failure, and the discovery of suitable techniques which could be employed to limit burning to certain surfaces of the propellant charge. Among the most significant advances was the development of methods to fabricate large propellant grains, of high quality, to precise geometrical shape.

Secondary developments of considerable importance were the development of suitable ignitor arrangements, methods of retaining propellant charges within the rocket motor, and techniques of exhaust nozzle and combustion chamber design. In addition, for artillery rockets, new fuse mechanisms for arming the explosive warhead had to be developed, since conventional shell fuses would not operate satisfactorily at low rocket accelerations.

The theory and semitheoretical correlation of factors important for both exterior and interior ballistics proceeded quite rapidly, and the fundamental understanding of rocket behavior and its dependence upon various parameters is now quite good. Rocketry grew from an art to a science, although much still remains to be done.

The central problem of interior ballistics is the combustion process of the solid propellant. Empirical relations, adequate for rocket design and analysis, were quickly established, but fundamental theoretical understanding is just now being reached [2,3,4,5] (see especially II,M). Additional problems, in the rocket motor, arise from the interaction of the gas flow resulting from combustion and the combustion process itself. The combustion mechanism is difficult to analyze because it is a two-phase process, the solid propellant transforms to a gas, and, because combustion is a rate process, the powerful tools of conventional thermodynamics are not applicable. Furthermore, the combination of chemical and gas dynamical features important in combustion make it a very complex process [6].

Exterior ballistic problems, very important for unguided missiles, are, in principle, rather more straightforward than problems of interior ballistics. These studies center around the problem of accuracy, and are concerned chiefly with the motion of a rigid body subject to aerodynamic forces and the rocket thrust force. As a matter of fact, the large number of factors that must be considered and the intricacy of the motion of a rocket missile has made these studies very difficult and at times has led to misunderstandings.

It is the purpose of the following discussion to sketch the main features of rocket theory and design. The large number of detailed considerations, quite important in any practical design, are mentioned.

H,2. Outline of Construction and Operation.

Shape. A rocket motor usually has the form of a circular cylinder, with length-to-diameter ratio considerably in excess of unity. In order to achieve a thrust exceeding its weight, and also good propellant utilization, the rocket must operate at internal pressures of several atmospheres. It follows that the combustion chamber must be a pressure vessel. Since it is desirable to have a large over-all impulse-to-weight ratio, the weight of the combustion chamber should be kept to a minimum. Stress considerations then indicate that spherical or circular cylindrical combustion chamber shape is desirable. A regular shape for the propellant charge is desirable, so that the burning surface area will not change widely during combustion. A large variation in burning surface area will produce extremely large variations in chamber pressure. The combustion chamber must then be designed to withstand a high peak combustion chamber pressure with a resultant increase in weight. Since it is much more feasible to design a regular propellant charge shape for the cylindrical combustion chamber than for a sphere, the cylindrical combustion chamber is usually preferred, despite a somewhat larger weight per contained volume than for a sphere. Rockets utilized to propel missiles usually have additional requirements of aerodynamic drag and stability, which make a cylindrical shape almost mandatory. Considerations of convenience in storage and transportation, and availability of seamless tubing, also make the cylindrical shape desirable.

Manner of operation. Before discussing solid propellant rockets further, it is desirable to describe their manner of operation and to outline the qualitative relationships which govern rocket performance.

The rocket unit consists of a charge of solid propellant within a combustion chamber and an exhaust nozzle through which the products of combustion of the solid propellant escape. The reaction due to the expulsion of the combustion products supplies the thrust of the rocket. The propellants used in rockets, just as those used in guns, do not explode but instead burn away at a definite rate on those surfaces which are exposed to the hot gases or flame within the combustion chamber. The burning rate depends upon the chamber pressure, increasing with higher pressures. Now, the thrust of a rocket motor may be considered as equal to the product of exhaust velocity and mass flow, so that, in order to get a large thrust, a large burning surface must be used to obtain a large mass flow. Similarly, to obtain a long duration of thrust only a small portion of the propellant charge must burn at a time. Since a combustion chamber of a given size contains only a certain weight of propellant, the thrust may be made large for a short time by providing a large burning surface, or small for a long time by providing a small burning surface.

The general pressure level within a rocket unit depends strongly on the ratio of the area of the burning surface to the cross-sectional area of the exhaust nozzle throat. For this reason, propellant-charge shapes are usually designed so that the area of the burning surface does not change extensively as the charge burns away. This requirement usually leads to regular geometric shapes for the propellant charge and to the arrangement of layers or strips of poorly combustible materials placed on the charge to prevent burning of certain areas (these layers are often called restrictors). In Fig. H,2a a variety of charge shapes which have been employed for rockets is shown.

The thrust of a rocket is proportional to the area of the burning surface, and depends upon the chamber pressure, the kind of propellant, and the internal geometry of the combustion chamber and propellant

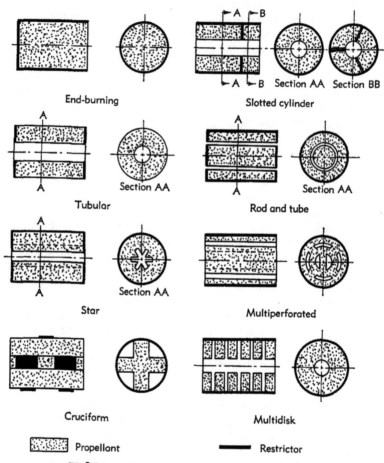

End-burning

Slotted cylinder   Section AA   Section BB

Tubular   Section AA

Rod and tube   Section AA

Star   Section AA

Multiperforated

Cruciform

Multidisk

▨ Propellant   ▬ Restrictor

Fig. H,2a. The geometry of solid propellant charges.

grain. The duration of thrust burning time is proportional to the web thickness of the propellant grain (the distance that the burning surface must recede in order to consume the propellant).

As mentioned above, the chamber pressure depends strongly on the ratio of the area of the burning surface to the exhaust nozzle discharge area, increasing rapidly as this ratio increases. The pressure also depends upon the internal geometry of the combustion chamber and the propellant charge.

Referring again to Fig. H,2a the simplest arrangement is the end-burning, or cigarette-burning geometry. Since a long grain may be used, the end-burning grain is an excellent method to obtain long duration of thrust. Durations from 1 to 500 seconds have been obtained in this way.

Internal-burning grains, star center and multiperforated in the examples shown, are becoming the most widely used. Although this geometry may lead to combustion stability problems, it minimizes the problem of combustion chamber heating. The grain itself insulates the combustion chamber wall until the final bits of propellant are consumed. In general, arrangements other than end-burning lead to complex problems associated with gas flow past the burning grain.

Practical developments. The solid propellant rocket, in its essential features, is remarkably simple. However, the development of a practical

Table H,2a. General characteristics of two types of solid propellant rocket motors.

Type	Radial-burning (tubular, cruciform, internal)	End-burning
Thrust	50 to 100,000 lb	10 to 5000 lb
Duration	0.05 to 50 sec	1 to 500 sec
Impulse	30 to 1,000,000 lb-sec	1000 to 100,000 lb-sec
Propellants commonly employed	Ballistite (extruded); Cordite (British, Russian, German); NDRC moulded composite; thermosetting, cast, composite; British plastic composite	Cordite; NDRC moulded composite; asphalt thermoplastic composite; synthetic rubber thermosetting composite
Applications	Artillery rockets, armor-piercing bombs, antiaircraft rockets, air to ground rockets, rockets for assisted take-off of light aircraft, glide bombs, suicide aircraft, booster rockets, long range rockets	Aircraft assisted take-off and superperformance, underwater projectiles, glide bombs, large airborne missiles, booster rockets, gas generation, sounding rockets
Jet blast*	Proportional to thrust, usually severe exhaust flame from ballistite	Proportional to thrust, long duration units usually have mild exhaust flame to reduce heating of mechanical components

* Most of the composite propellants produce a smoky exhaust jet.

rocket motor that is reliable, convenient to use, and moderately easy to manufacture has been a lengthy and complex process. Because of the importance of these practical considerations, discussion of solid propellant rockets must necessarily refer to the state of the art.

In Table H,2a an outline is given of the general scope of performance and application attained by solid propellant rockets. Rapid development makes it difficult to separate experimental units or propellants from successful utilization, and it also makes a table of this type subject to rapid obsolescence. However, as a rough general survey, it should prove helpful.

The flexibility in design that is both useful and possible is very great. By utilization of suitable designs for various applications it is possible to obtain thrust variations and burning time variations ranging over a factor of 10^4. Total impulse, in different designs, can vary by factors greater than 10^6. Approximately 14 different applications and 10 different propellants are listed; both applications and propellants represent main lines of effort, and the number of each would increase considerably if all the various modifications and models were taken into account.

It will be noted that the ratio of total impulse to total loaded weight has not varied over such a wide range, due to the limitation on the energy per unit mass that can be obtained from solid chemical fuels. Because of this, features of rocket performance, other than specific impulse, receive considerable attention.

Table H,2b. Characteristics of some solid propellants.*

Property	Ballistite extruded JPN	Asphalt base thermoplastic composite	NDRC moulded composite
Specific impulse, sec	220	186	170
Exhaust velocity, ft/sec	7100	6000	5500
Density, lb/ft³	101.5	110	100 to 115
Specific impulse/unit volume, per cent	100	96	80 to 90
Burning rate, in./sec	1.4	1.6	0.25 to 1.32
Burning rate exponent	0.73	0.76	0.40
Area ratio†	185	175	200 to 1000
Low pressure combustion limit, lb/in.²	500	1000	no lower limit
Flame temperature, °F	5300	3000	3000
Exhaust jet	smokeless	smoky	smoky

* All pressure-dependent properties assume good exhaust nozzle design, 2000 lb/in.² abs combustion pressure, and 14.7 lb/in.² abs external pressure. The combustion process is temperature-sensitive, and 70°F has been assumed.

† Ratio of burning surface area to exhaust nozzle throat area.

Table H,2b lists the characteristics of some solid propellants, which serve to provide orientation concerning propellant performance. The variation in available specific impulse amounts to only 25 per cent. Burning rates, at 2000 lb/in.² abs, are all less than 2 in./sec. It is possible to achieve low burning rates, but very high burning rates, for example 20 in./sec, are not available. Such high burning rates could be very useful in certain applications.

Table H,2c describes an artillery rocket and an assisted take-off rocket. The examples chosen do not represent either the largest or the smallest of their kind, and very many other rockets could be cited.

The curves in Fig. H,2b show the variation of thrust and chamber pressure for a large, end-burning, asphaltic composite propellant rocket. The regularity and smooth operation of this long duration (23-sec) unit is evident.

In Fig. H,2c the curves show the variation of chamber pressure with time for a cruciform propellant charge. The difference in performance for different initial propellant charge temperatures is noteworthy. This illustrates one of the fundamental problems in rocket design,

Table H,2c. Characteristics of two rockets.

Rocket description	3.5 in. Fin-stabilized	12 AS-1000 D-1
Charge type	Cruciform	End-burning
Purpose	Artillery	Assisted take-off of aircraft
Thrust, lb	2340	1000 (rated)
Burning time, sec	0.8 sec at 70°F	12 (rated)
Impulse, lb-sec	1870	16,000 (actual)
Propellant	JPN	Asphaltic composite
Outside tube diameter, in.	3.5	9.6
Length, in.	54.5	36
Gross weight, lb	53.8	205
Propellant weight, lb	8.5	90
Payload weight, lb	20.0	—
Launcher weight, lb	15 (air)	—
Range, yd	9000*	—
Burnt velocity, ft/sec	1175	—
Dispersion, mils		
Air-fired	4	—
Ground-fired	20	—

* Estimated.

temperature sensitivity. Except at the extreme upper propellant temperature, it will be noted that these performance curves are very regular.

There are three characteristics, peculiar to solid propellant rockets, which limit their utilization. These limitations may be briefly designated by the following terms:

1. Temperature sensitivity—the variation of performance with propellant charge temperature.
2. Temperature limits. A rocket unit will malfunction if its temperature, before firing, does not lie between certain limiting temperatures.
3. The combustion limit. Many propellants do not burn satisfactorily at low pressures.

Propulsion by detonation. From time to time, suggestions are put forward that a "rocket" should be propelled by detonation. The advent of the shaped charge with particle ejection velocities of 30,000 ft/sec has given further impetus to these suggestions. There are, of course, some practical objections to such an arrangement. The violence of detonation

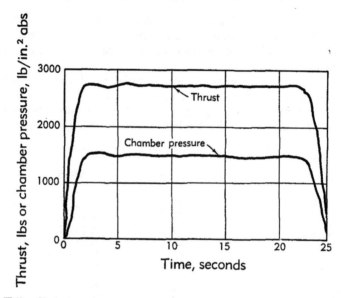

Fig. H,2b. Variation of thrust and chamber pressure with time for a large, end-burning, thermoplastic, asphaltic, solid propellant rocket. Performance data: $p_c = 1465$ lb/in.^{2} abs, $F = 2660$ lb, $I = 62,912$ lb sec, $W_o = 349.5$ lb, $I_{sp} = 180$ sec., $A_{th} = 1.179$ in.^{2}, $C_F = 1.54$.

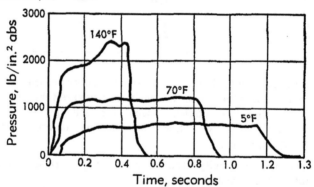

Fig. H,2c. The variation of chamber pressure with time for a 9-lb cruciform, semi-restricted-burning, ballistite grain for the 3.25-in. rocket motor. Note the effect of propellant charge temperature on performance.

would be very destructive to the launcher, and probably destructive to the rocket itself. In the case of the shaped charge, it has been shown [7] that the high speed penetrating jet is balanced by a much larger low speed jet emitted in the opposite direction. A feasible method to utilize the low speed jet to "push" a vehicle has yet to be demonstrated.

However, there is a more fundamental objection. Any jet propulsion device is limited by the conservation laws: conservation of energy, mass, and momentum. The detonating material will be superior to the common rocket propellants only if it can release more energy per unit mass than the latter. This is not generally the case. The improved explosives achieve their greater efficacy by having a greater rate of detonation, and the energy per unit mass is about the same as in the case of propellants. Indeed, ballistite has a very high specific energy and few solid materials exceed it in energy per unit mass.

It may be argued that the pressure level in a detonation is much higher than in the case of rocket motors and that, therefore, a thermodynamic gain in efficiency is to be looked for. However, the pressures commonly utilized in rocket motors usually are high enough so that a gain in exhaust velocity of 5 or 10 per cent is the most that can be expected on going to very high pressures.

The only remaining advantage would seem to be the possibility of achieving a different distribution of energy over the ejected mass rather than the uniform distribution obtained with rocket motors. Consider an abstract case: Suppose one has a mass M and a total amount of energy E. What is the optimum distribution of energy over this mass in order to achieve the maximum impulse from a rocket device? The corresponding variational problem seeks to maximize the total impulse $I = \int V_e(m)dm$, where m is the differential mass coordinate such that $\int dm = M$, with the constraint that the total energy is constant, $E = \frac{1}{2}\int V_e^2 dm = $ const. The solution of this problem states that the maximum impulse is obtained when every particle, dm, is ejected with the same exhaust velocity $V_e^2 = 2E/M$.

In conclusion, one may say that from a fundamental point of view there are at best very minor advantages in going to a detonating explosive as a means of propulsion, and that there are very great practical difficulties in the way of such a development.

H,3. Effect of Utilization on Rocket Design. The fundamental purpose of a rocket unit is to supply a required amount of impulse, force × time or mass × velocity, to a vehicle. Rockets are not usually employed if the fundamental purpose is to do work (move a force through a distance) but rather, from a general point of view, they are used to accelerate a mass to some velocity. The final velocity should be moderately high (an appreciable fraction of the exhaust velocity of the rocket) if the rocket is to be used efficiently. However, in many cases, convenience rather than efficiency is the feature of the rocket which recommends it for a particular application.

Rockets have been utilized to assist the take-off of an aircraft, to supply an initial boost to a missile, and to supply the main propulsive force for a missile. A modification of the rocket arrangement, in which the high pressure gas resulting from combustion is utilized to operate pneumatic devices, has also been employed. This last purpose, gas generation, is at complete variance with the missile applications mentioned above, but it has grown as a by-product of rocket development as the reliability of rocket devices has increased.

General requirements. Safety and reliability, with adequate performance, are nearly always demanded from rockets. Sensitive or unstable propellants must be avoided, even though their use might improve certain aspects of rocket performance. Safety also limits rocket ignitor design and the composition of ignition powder. The electric squib commonly used with ignitors usually requires a minimum of 0.5 amperes (resistance of about 1 ohm implies 0.5 volt) for ignition to minimize accidental firing. Squibs with smaller energy requirements can be obtained and would place a smaller load on firing circuits. However, use of such squibs would increase the danger of accidental firing.

The ignitor squib ignites the ignitor powder. Here, there are advantages, reduced ignition time and reduced ignitor blast, which may be gained by employing special powders.

Since rockets are self-propulsive, they can constitute a hazard to a wide area if accidentally set off. They are frequently shipped in a partially disassembled, nonpropulsive condition to minimize this danger. In handling rockets it is well to remember that both the front (missile) and nozzle (blast) ends are dangerous.

The basic simplicity of solid propellant rockets has made it possible to attain a high degree of reliability with them. There remain limits in the range of conditions under which they must be stored and used if reliable behavior is to be obtained. The most important requirement is usually that suitable temperatures, neither too high nor too low, be maintained. Rugged design and resistance to shock and vibration are required, and this requirement can usually be met.

The attainment of reproducible performance has been a slow and difficult part of rocket development. Even now, what has been achieved is essentially identical performance under given conditions (the most important condition is rocket temperature before firing), but not identical performance under all conditions. Reproducibility requires meticulous control of propellant composition and processing.

A rocket must meet certain performance requirements. For the rocket itself, the most natural specification is total impulse. Independent specification of thrust and duration of thrust is often necessary. Since a given weight of propellant will deliver, rather closely, the same impulse under a wide variety of conditions (although the burning time and thrust may vary oppositely over a wide range), specifications frequently take the form of a required impulse with an additional requirement of minimum and maximum thrust or else maximum burning time. Separate thrust and duration requirements then serve only to define the area of interest.

Rockets to assist the take-off of aircraft usually have large impulse but moderate thrust (to avoid large, sudden loads on the aircraft structure) requirements (12,000 lb-sec, 1000 lb), and the long burning time is most easily attained by end-burning grains. For missile boosters, a minimum thrust may be specified (to get the missile up to a speed at which aerodynamic control is effective). For reasons of accuracy, artillery rockets must attain a certain speed on the launcher, and a maximum burning time (which implies a minimum thrust) may be required. Rockets for boosters and missiles usually require such a large impulse and short burning time that internal-or radial-burning grains are used.

Prompt ignition is nearly always desirable. For assisted take-off of aircraft, ignition delays up to $\frac{1}{2}$ second are tolerable. For missiles, which may be fired at moving targets or in ripple salvos, delays should not exceed 10 to 50 milliseconds (depending on the application). The sustaining rocket in a boosted missile should start within a similar time interval, since its start-up must synchronize with the release of the booster.

Requiring short ignition delay usually implies very energetic ignitors, which, in turn, may damage the propellant charge, and even the structures at the launching site to the rear of the rocket. The ignitor blast may be more severe than the rocket blast (exhaust jet). Fragments of the ignitor (sometimes burning) may be ejected through the exhaust nozzle, thus constituting a hazard.

Ideally the exhaust jet of the rocket should not be visible (smoke, flash), and should not contain toxic or corrosive chemicals. Most rocket exhausts do contain carbon monoxide. Smoke is particularly undesirable for the assisted take-off of shipboard aircraft, since it may obscure the flight deck.

High velocity rocket missiles usually require a large ratio of total impulse to total weight together with a small frontal area, and a burning time of about one second. Internal-burning, traverse-burning, or radial-burning grains must be used, and the high internal gas flow velocity resulting from the large impulse and small frontal area requirements may give rise to combustion stability problems. The long grain is also difficult to support during the entire burning period, and may break up before delivering its full impulse.

Very short burning time rockets, designed to burn completely on the launcher, are subject to similar difficulties.

The proper balance between propellant performance and cost is always a difficult problem. When very large impulse is required (aircraft assisted take-off, large boosters), cost usually is given increasing consideration.

Artillery rockets. In addition to the usual requirements outlined above, the problem of accuracy introduces additional requirements and refinements for artillery rockets. The accuracy problem of exterior ballistics has been analyzed in great detail by Davis, Blitzer, and Follin [8], and by Rosser, Newton, and Gross [9]. Only the implications for rocket design are considered here.

Since range is closely correlated with impulse-to-weight ratio, both the propellant weight and total rocket weight must be controlled.

For most artillery rockets, only a small fraction of the rocket motor impulse is delivered while the rocket is on the launcher. For rockets fired from stationary launchers, more than 90 per cent of the dispersion (inaccuracy) may be attributed to motion during the burning period. This implies that proper alignment of the exhaust jet is important to a much larger degree than for booster rockets for a guided missile. As a rough order of magnitude, for a missile stabilized by external fins, if a line along the direction of thrust misses the center of mass of the rocket by 0.10 inch, a lateral deviation of 20 to 30 mils (1 mil is an angle corresponding to one yard at 1000 yards) is produced.

Thrust malalignment has been attributed to the following causes: manufacturing tolerance, gas malalignment, and thermal and pressure distortion of the rocket body.

Gas malalignment occurs when the exhaust jet does not coincide with the axis of a mechanically perfect exhaust nozzle. There is some indication that a high turbulence level in the gas flow reduces gas malalignment. Thermal and pressure distortion of the combustion chamber, if nonuniform, can make one side longer than the other, canting the exhaust nozzle to one side. For example, a 60°C mean temperature difference on two sides of a steel combustion chamber 3 feet long and 3 inches in diameter, can produce thrust malalignment of 0.06 inches (corresponds to 10- to 20-mil deviation). Insulation of the combustion chamber wall can reduce thermal distortion. To minimize pressure distortion, all parts must expand equally; closures, lugs, and fittings must be designed with this in mind, and the combustion chamber wall should have uniform thickness.

The effect of thrust malalignment may be reduced by spinning the rocket about its longitudinal axis. This is automatic with spin-stabilized rockets, and even fin-stabilized rockets can be given a slow spin.

Multiple exhaust nozzles may also be used to reduce thrust malalignment. The individual imperfections neutralize statistically.

High velocity fin-stabilized rockets may have a length-to-diameter ratio of 10 to 14 with attendant interior ballistics problems. Spin-stabilized rockets with a length-to-diameter ratio of 5 to 8 have easier interior ballistic problems.

The ignitor may contribute to inaccuracy if a portion of the ignitor, passing through the exhaust nozzle, disturbs the exhaust jet.

The support of the propellant charge throughout the burning period is also important. Break-up of the grain at the later stages of burning produces slivers of propellant which can disturb the exhaust jet on passing through the nozzle, and also represent a loss of impulse (range). Cruciform charge shapes, rather than tubular, were a first step to provide better charge support. Internal-burning grains now appear to be an even better solution (Art. 14).

CHAPTER 2. INTERIOR BALLISTICS THEORY

H,4. Scope of the Theory. The theory presented here is idealized in that perfect gas laws, simple burning rate behavior, ideal thermodynamic propellant characteristics, and steady state flows are assumed. With this simplified framework, analytic expressions may be derived which give more insight into the basic features of rocket behavior than the more precise numerical techniques. In many instances, the theory can be made quantitatively useful by semiempirical choice of constants; indeed, the burning rate expressions that are employed are essentially empirical in character.

The simplified theory is also extrapolated to very high pressures where it has only a schematic meaning. However, by doing this, one may obtain a broad survey of rocket processes. Most of the qualitatively different phenomena associated with steady state processes may be studied in this way.

Unsteady state processes, which can often be of great practical importance, are not treated here. Resonance burning, a combustion instability problem of rocket design, is not analyzed. The theory is complex and still controversial [*10; 11; 12*, pp. 893–906; *13*, pp. 29–40; *14; 15; 16*] (also II,M). Certain practical design arrangements to avoid this difficulty are listed below in Art. 10.

In addition to providing insight, the theory provides a helpful framework and useful analytical expressions for the semiempirical analysis of rocket data. A more comprehensive account of some of these problems is given by Crawford [*17*], Wimpress [*10*], and Price [*18,19,20,21*]. Very brief accounts are given by Sutton [*22*] and Seifert, Mills, and Summerfield [*23*].

H,5. Combustion of Solid Propellants. The general manner of operation of a rocket unit has been outlined. In most solid propellant rockets, no arrangement is made for the control of the combustion process during the burning of the propellant. If there are no mechanical valves or other devices installed in the unit (aside from certain safety devices) by which the burning of the propellant may be regulated, the control of the burning of the propellant is possible only to the extent to which a suitable grain and combustion chamber design, along with some arrangement of inhibitor coatings, is prepared before the unit is used. After ignition, the unit is "left to itself" and usually no further adjustment on the part of the operator is attempted. Provision for special control of the unit is a difficult design problem and could make the rocket unit undesirably complicated. One of the chief attractions of solid propellant rockets has been their simplicity and reliability.[1]

During the burning period, the solid propellant rocket regulates itself. The fundamental property of the solid propellant involved in this self-control of the burning process is its rate of burning. The rate of burning of a solid propellant is defined as the rate at which a burning surface recedes in a direction normal to itself. This is usually expressed as a velocity, e.g. in./sec.

Experimentally, burning rates can be determined in static firing tests of the rocket units or by strand burners (see Art. 12). If the geometry of the propellant charge is known, the burning time can usually be estimated from the pressure vs. time curve obtained in the test. From this information the rate of burning of the propellant may be calculated.

From special experimental studies it has been found that the burning rate r depends upon: the gas pressure on the propellant p, the temperature of the propellant charge before ignition T_p, the velocity of gas flow past the charge surface V, the time since the start of burning t, the position in the charge x, and the direction in which the normal to the burning surface points Ω. In mathematical form this may be expressed as

$$r = r(p, T_p, V, t, x, \Omega)$$

It is important to remember that all these parameters can affect the burning rate. However, not all of them are important for every propellant.

The variation in burning rate with time after start of combustion is believed to be due to gradual heating of the propellant and motor tube by radiant heat transfer from the hot propellant gas. Avery [24,25] has pointed this out, and a summary of studies of this problem has been prepared by Penner [26]. It is a more important effect for hot propellants such as ballistite and cordite than for cool propellants. The propellant grain must be slightly transparent to radiation to permit internal heating. (Very transparent propellants are subject to subsurface ignition by radiation, which must be avoided. See Art. 12.)

---

[1] Sometimes a spring-loaded valve discharging into an auxiliary exhaust nozzle is employed with rockets. The effect of this is to allow the mass outflow to increase with pressure more rapidly than the linear variation which is obtained with ordinary nozzles. This improves the stability of operation, but it does not constitute independent "external" regulation.

Both the variation with position and direction are presumably due to inhomogeneities and anisotropies introduced into the structure of the propellant during its manufacture. For solventless ballistite, extruded in the form of a tube, the burning rate may be 25 per cent higher at the center of the web than at the surface [10]. (However, this may be due to the radiant effects mentioned above.) The burning rate is also some 15 per cent higher in the direction of extrusion than it is at right angles. Cast, polymerized propellants utilized in end-burning charges show a higher burning rate near the mold wall (Art. 12). The asphaltic, thermoplastic, cast propellants do not seem to show the dependence of burning rate on time, position, or direction.

Although the detailed burning rate behavior of a propellant grain must be known in order to reliably design a rocket motor, it will simplify the general theory if only the variables of chamber pressure, propellant temperature, and gas flow velocity are retained. There are indications [10,27,28,29,30,31] (see also II,M) that the effect of temperature is separable from the pressure and velocity dependence:

$$r = f_1(T_p)f_2(p, V)$$

Experimental studies on the burning rate of propellants show that the temperature dependence can be expressed by

$$f_1(T_p) = \frac{A}{T_1 - T_p}$$

Here A and T_1 are empirical constants.

The pressure and velocity dependence seem to be separable in some cases [7], but not in others. Two forms are commonly employed [32,33]:

$$f_2(p_o, V) = a'\left(\frac{p}{1000}\right)^n (1 + k\rho V) \quad \text{or} \quad a'\left(\frac{p}{1000}\right)^n f_3(V)$$

Here the new parameter ρ is the density of propellant gas flowing past the grain, and a' and k are empirical constants. The density may be expressed in terms of pressure and velocity so that the first expression is consistent with f_2.

In the second form, $f_3(V)$ has a peculiar dependence:

$$f_3(V) = \begin{cases} 1 & \text{if } V < V_0 \\ 1 + k_V(V - V_0) & \text{if } V > V_0 \end{cases} \tag{5-1}$$

Here k_V and V_0 are empirical constants, and V_0 is a threshold velocity. Frequently an expression without a threshold velocity is employed:

$$f_3(V) = 1 + kV$$

The threshold situation is outlined in Fig. H,5a. The experimental curve is fitted rather well by Eq. 5-1 over most of the velocity region of interest.

The experimental results indicate that no acceleration of the burning rate takes place (referred to as erosive burning) until a velocity of 600 ft/sec is reached. This may be due to the buffer effect of propellant gas coming directly from the surface.

The factor $1 + k\rho V$, on the other hand, has more intuitive appeal, since transfer of mass, momentum, and heat across a boundary layer often show this dependence on ρV. The dimensional erosion constant k usually ranges from about 0.05 to 0.7 in.² sec/lb. The theory of solid

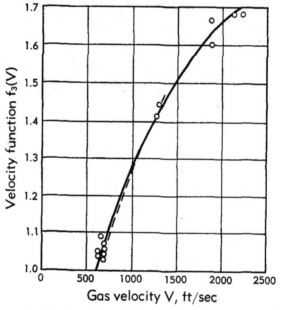

Fig. H,5a.   Erosive burning effect, velocity-dependent only, of gas velocity on the burning rate of JPN propellant. After [10] by permission.

propellant burning without gas flow past the surface [30,34,35,36,37,38] is still being formulated. Both the theory [6] and facts [32,33,39,40,41] of erosive burning are still unsettled. Corner [6] indicates the theoretical possibility for either velocity or mass flow density erosive effects depending upon the propellant. The latter should usually be accompanied by an index n in the pressure dependence of 0.5 or less. The variation of k or k_V with propellant temperature is also not settled. Here they are assumed independent of propellant temperature.

Green [38] has carefully studied erosive burning effects (dependence of burning rate on gas flow velocity is frequently termed erosion), and he suggests that the peculiar threshold effect (Eq. 5-1) may be due to the shape of the gas flow velocity profile at the forward end of the propellant grain, and that the intrinsic velocity dependence is of the form $1 + kV$.

He was unable to decide conclusively whether or not the form should be $1 + kV$ or $1 + k\rho V$, although experiments performed by Bartley indicate conclusively that the latter expression is more suitable for certain composite propellants.

Green also suggests that the proper function should not be velocity but the dimensionless mass flow density (Art. 8). This would correspond to an erosive effect independent of density.

In the following discussion, both mass flow density $1 + k\rho V$ and velocity $1 + kV$ erosion effects will be included. The former has some experimental basis and also gives rise to an erosive instability which can be of practical importance for certain propellants. The "threshold" form (Eq. 5-1) will be chosen for the velocity-dependent erosion, since, according to Green, this is an empirical method to take into account the two-dimensional flow at the forward end of the grain. It also provides a qualitative comparison between the effect of gas flow on erosive and on nonerosive propellants.

The burning rate expressions adopted here may then be written

$$r = \frac{a'}{T_1 - T_p} \left(\frac{p}{1000}\right)^n f_3(V) \qquad (5\text{-}2a)$$

$$r = \frac{a'}{T_1 - T_p} \left(\frac{p}{1000}\right)^n [1 + k\rho V] \qquad (5\text{-}2b)$$

For negligible velocity these reduce to

$$r = a \left(\frac{p}{1000}\right)^n \qquad (5\text{-}3)$$

with

$$a = \frac{a'}{(T_1 - T_p)}$$

The pressure is to be expressed in psia and the parameter a has values ranging from 0.1 to 1.0 in./sec. Values of n are less than unity, frequently ranging from 0.2 to 0.8 (Table H,2b and Art. 12).

In Eq. 5-2 the form chosen for temperature dependence is suggestive, indicating an infinite burning rate at $T_p = T_1$. Efforts to identify T_1 with an ignition temperature, T_i, have not always been quantitatively successful. For JP ballistite the ignition temperature was determined to be 330°F [42], while the constant T_1 determined from temperature sensitivity studied was found to be 440°F. However, smaller values of T_1 do indicate a larger degree of temperature sensitivity.

Special propellants have been devised which exhibit a constant burning rate, independent of pressure, over certain pressure ranges. Such propellants are particularly desirable since they yield extremely stable

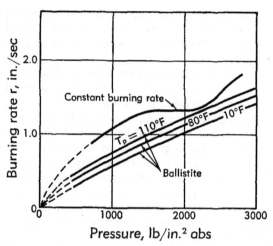

Fig. H,5b. Variation of burning rate with propellant gas pressure and propellant charge temperature for ballistite (JPN), together with an example of "constant burning rate" propellant.

rocket unit behavior. In Fig. H,5b burning rate curves for ballistite and a curve for such a constant burning rate propellant are shown.

H,6. Stability of the Burning Surface.

Combustion geometry. The solid propellant burning surface, in many cases, moves in a direction normal to itself at a uniform rate at every point. Propellants which are not macroscopically homogeneous and isotropic exhibit a different, although regular, behavior. It is important that the regular behavior of the burning surface should be stable, so that the burning surface area does not change in an unpredictable manner. If a disturbance is represented by an irregularity of the surface (this might be caused by a fault or large inhomogeneity in the propellant) simple geometrical considerations, assuming burning in a normal direction, indicate that the surface will smooth out and continue its regular recession as long as the desired surface is flat. Fig. H,6a shows this for an end-burning propellant charge. The burning surface behavior predicted for the cup-shaped burning surface (second from the top) has been verified experimentally. These same general considerations are valid for charges of other shapes as long as the disturbance is small compared to the radius of curvature of the charge surface. On the other hand, a large disturbance, on a cylindrical surface for example, will grow although the area of the burning surface will decrease slightly so that dangerously high pressures are not developed. A constant area for the burning surface is the important requirement, since a small change in the burning surface area will produce a large change in the chamber pressure. One may conclude that the burning surface is stable (see also Art. 12).

Charge deformation problems. The propellant charge is subject to pressure, and in many cases axial and radial acceleration forces. These forces produce deformation of the charge. In the case of end-burning propellants, sealed to the combustion chamber walls, the application of chamber

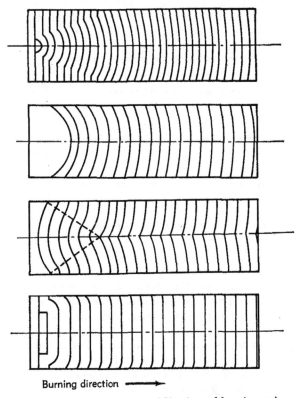

Burning direction ➤

Fig. H,6a. Burning surface stability for end-burning grains.

pressure can lead to a reduction in charge length of 5 to 10 per cent. If the charge is contained within a flexible casing, not sealed to the combustion chamber wall, then the build-up of chamber pressure will apply a hydrostatic load to the propellant with little deformation and very little volume reduction since the Poisson ratio is near $\frac{1}{2}$ for many propellant materials.

In the case of tubular and internal-burning propellants, in which gas must flow through ducts along the grain, the situation is more serious. The general pressure level does not lead to a deformation of the charge, but the chamber pressure at the forward end of the rocket is larger than the chamber pressure at the rear or exhaust nozzle end of the charge.

The unbalance of pressure does lead to a nonisotropic, longitudinal compressive stress, bulging the charge into the gas duct area. For artillery rockets there are large acceleration forces which have the same effect. Skin drag and any impact forces on propellant irregularities from the rapidly flowing gas stream also add to the longitudinal compression, as illustrated in Fig. H,6b.

The bulging of the propellant charge into the gas channel leads to a restriction which further increases the pressure drop, etc. At high propellant temperatures the burning rate is higher, the propellant is softer,

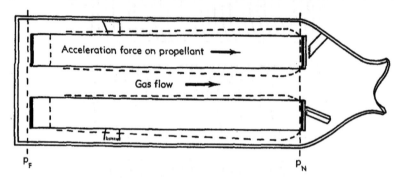

Fig. H,6b. Deformation of the propellant by pressure drop, gas drag, and rocket acceleration forces. The solid line represents the undistorted charge; the dashed line represents the distorted charge.

the normal duct area is reduced by thermal expansion of the propellant (which is much larger than the expansion of the combustion chamber). The higher pressure produces larger accelerations. All these effects result in increased pressure drop and bulging, and may lead to failure of the rocket.

This problem can be worked out in detail. For illustration, acceleration and skin friction forces are omitted and only the pressure drop due to the necessary gas flow velocity is considered. At the critical time, initiation of burning, frictional forces are small.

If the cross-sectional area of the propellant is normally A_{pn}, then the effect of pressure drop is to increase this to A_p, where

$$A_p = A_{pn}\left(1 + \frac{2\mu}{E}\sigma\right) \tag{6-1}$$

$$\sigma = p_F - P_N \tag{6-2}$$

The front pressure is p_F, the rear pressure p_N (Fig. H,6b), and the stress component leading to deformation is therefore σ. Poisson's ratio and the elastic modulus are indicated by μ and E. If the internal cross-sectional area of the combustion chamber tube is A_q, then the duct area

⟨ 84 ⟩

A_d is

$$A_d = A_q - A_p, \quad A_{dn} = A_q - A_{pn} \qquad (6\text{-}3)$$

The subscript n corresponds to the normal (unbulged) state. For illustrative purposes we assume that $A_q = 2A_{pn}$, a value not far from those employed in rockets. This will simplify the algebra, but it will not affect the fundamental character of the results. If the propellant were perfectly rigid, the pressure drop would be less, σ_n, since no bulge would obstruct

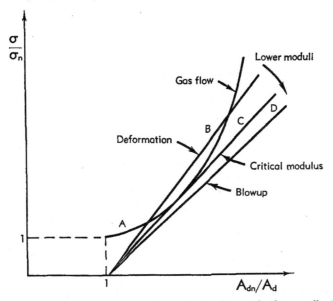

Fig. H,6c. Schematic curves showing the variation of stress in the propellant grain with gas duct area A_d. The curved line is the stress due to gas flow, the nearly straight lines represent the stress required to maintain the corresponding gas duct area. The latter curves rotate clockwise for lower values of the elastic modulus, and the curve C corresponds to the critical modulus.

the duct area. Eq. 6-1 and 6-2 may be combined to give

$$\frac{\sigma}{\sigma_n} = \frac{1}{2\mu} \frac{E}{\sigma_n} \left[1 - \frac{1}{(A_{dn}/A_d)} \right] \qquad (6\text{-}4)$$

It is necessary to use a result, which is derived below (Art. 9), to complete the analysis. This is the relation between pressure drop σ and the gas duct area when the remaining parameters of the rocket (the burning surface area S_b and the exhaust nozzle throat area A_t) remain fixed. This result, together with the result of Eq. 6-4, is shown in Fig. H,6c. The curve labeled "gas flow" follows from the interior ballistics study (Art. 9), and the curve labeled "deformation" is the result of Eq. 6-4. An important deduction from this study is that there is a critical elastic modulus, and if the actual modulus of the propellant is less than this the rocket will blow up.

Consider the intersection point A. If the duct area were to decrease slightly by a disturbance (this corresponds to a motion to the right of A since the duct area is in the denominator of the horizontal coordinate), then the stress due to gas flow will be less than the elastic stress required to maintain the reduced duct area. Consequently the compression will relax and the conditions return to the intersection point. Similarly a disturbance causing a displacement to the left of A is followed by a decrease in duct area until A is reached. A is the stable intersection point and corresponds to the actual duct area and the actual compressive stress when charge deformation is taken into account. From the stress and Eq. 6-2 the actual chamber pressure distribution may be determined.

A disturbance, leading to a displacement to the right of the intersection point B, yields a higher compressive stress due to gas flow than is required to maintain the reduced duct area. The duct is reduced further, the pressure increases further, and the rocket blows up. Below B, pressures decrease until A is reached. B is an unstable equilibrium point.

Inspection of Eq. 6-4 shows that a reduction in the elastic modulus rotates the "deformation" curve clockwise, bringing the points of intersection of A and B closer together. If these points come too close together, a disturbance of the true equilibrium conditions at A might throw the system to the right of B and lead to a blowup. Finally, further reduction of the modulus leads to curve C, just tangent to the gas flow curve. An ideal rocket might operate here, but a slight disturbance could lead to a blowup. Further reduction in the elastic modulus of the propellant would remove all points of intersection of the "gas flow" and "deformation" curves (see curve D). The compressive stress from gas flow would always exceed the stress necessary to maintain the reduced duct area, leading to further reduction in duct area and further increase in pressure until the rocket blows up. The modulus, corresponding to curve C, is the critical modulus. A rocket should be proportioned so that the critical elastic modulus is substantially less than the actual modulus. The critical modulus depends only upon the pressure drop, duct area characteristics of the unit, and is determined by the geometry of the arrangement and the burning characteristics of the propellant.

It was mentioned above that flight acceleration and increased propellant temperature tend to increase the degree of propellant deformation. Wimpress [10] has computed curves taking all these effects into account, and his results are shown in Fig. H,6d. The intersection of the actual moduli curves with the critical modulus curve indicates the upper temperature limit (blowup) for the rocket. Agreement between theoretical upper temperature limits determined from these curves and firing tests is reasonably good. It should be noted that reduction in actual modulus

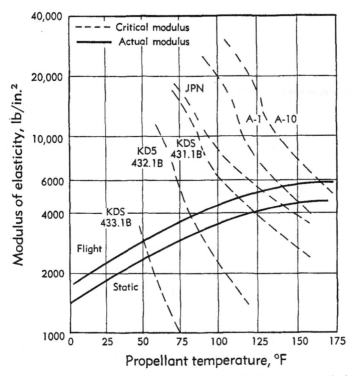

Fig. H,6d. Variation of the critical modulus for the Mkl3 propellant grain in the 3.25-in. rocket motor. The variation of the actual elastic modulae for a number of propellants with propellant temperature is shown on the same scale. The intersection "critical-actual" corresponds to the upper temperature limit (after Wimpress [10] by permission).

with temperature is a more important effect than increase of critical elastic modulus with temperature.

H,7. Steady State Dynamics for End-Burning Grains. The theory [43] and stability criteria for the end-burning charge are relatively clear and simple, and they are discussed first. The notation that is used is indicated in Fig. H,7a.

It will be shown, provided that the exponential index n in the burning rate expression, Eq. 5-3, is equal to or less than unity and the burning surface area is constant, that a stable, steady chamber pressure is obtained during the combustion of the propellant. In a practical rocket, the index should be appreciably less than unity. It is assumed that the temperature of the combustion gas is T_0, a constant. An analysis of transient startup conditions indicates that in less than 0.04 sec after ignition, the cool gas, usually air, initially present in the combustion chamber has been swept out.

The equation for the conservation of mass is simply:

$$\left\{\begin{matrix} \text{Mass} \\ \text{combustion} \\ \text{rate} \end{matrix}\right\} = \left\{\begin{matrix} \text{Rate of increase of the} \\ \text{mass of combustion products} \\ \text{inside combustion chamber} \end{matrix}\right\} + \left\{\begin{matrix} \text{Rate of mass} \\ \text{flow out} \\ \text{exhaust nozzle} \end{matrix}\right\}$$

In symbolic form,

$$r S_b \rho_p = \frac{d}{dt}(\rho_c v_c) + \frac{p_c A_t}{c^*} \tag{7-1}$$

Here r is the burning rate, S_b the area of the burning surface assumed constant, ρ_c the density of combustion gas in the combustion chamber, ρ_p the density of the solid propellant before combustion, v_c the volume

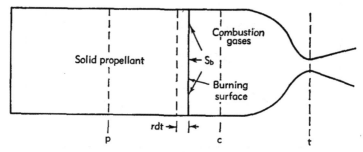

Fig. H,7a. Notation employed in the discussion of
end-burning solid propellant grains.

of the combustion chamber from the burning surface to the exhaust nozzle throat section, p_c the pressure within the combustion chamber, A_t the exhaust nozzle throat area, and c^* the characteristic velocity for the products of combustion[2] (see Sec. G).

The ratio of nozzle throat area to burning surface area is so small, in end-burning grains, that the approach velocity of the combustion gas inside the combustion chamber is negligible, so that p_c may be employed for the stagnation pressure or reservoir pressure in computing nozzle mass flow.

The products of combustion are treated as ideal gases so that the

[2] The exhaust nozzle flow equations are not developed here (see Sec. G). Often a discharge coefficient C_D is employed so that the mass flow $\dot{m}$ is given by $\dot{m} = C_D p_c A_t$. The coefficient, as well as c^* have a theoretical basis, but may also be determined experimentally. For real propellants they depend weakly on chamber pressure; this is neglected here. For ideal gases the relation between the velocity of the gas at the exhaust nozzle throat V_t ($= a_t$), the speed of sound in the combustion chamber a_c, and c^* is given by $c^* = a_c / \sqrt{\gamma} = \dfrac{V_t}{\sqrt{\dfrac{2\gamma}{\gamma+1}}} \Gamma$. Γ is defined in Eq. 7-5c. Note that the

form of Eq. 7-4 implies the use of absolute units. For engineering units, c^* should be replaced by c^*/g which has the physical dimensions of sec.

equation of state may be employed:

$$p_o = \rho_o \mathfrak{R} T_0 \qquad (7\text{-}2)$$

where $\mathfrak{R}$ is the engineering gas content (containing the average molecular weight) and T_0 is the adiabatic, constant pressure, flame temperature for the propellant. It is also convenient to introduce the abbreviation,[3]

$$p_p = \rho_p \mathfrak{R} T_0 \qquad (7\text{-}3)$$

If Eq. 7-1 is multiplied through by $\mathfrak{R} T_0$, and Eq. 7-2 and 7-3 employed,

$$r S_b p_p = \frac{d}{dt} (v_o p_o) + \frac{\mathfrak{R} T_0 A_t}{c^*} p_o \qquad (7\text{-}4)$$

The T_0 may be taken inside the differentiation sign since it is assumed constant.

It is convenient to introduce the ideal expression for c^*

$$c^{*2} = \frac{\mathfrak{R} T_0}{\Gamma^2} \qquad (7\text{-}5a)$$

$$\Gamma^2 = \gamma \left(\frac{2}{\gamma + 1} \right)^{\frac{\gamma+1}{\gamma-1}} \qquad (7\text{-}5b)$$

$$a_c = \sqrt{\gamma} \, \Gamma c^* \qquad (7\text{-}5c)$$

where γ is the ratio of specific heats—constant pressure to constant volume. For most rocket gases γ is near 1.2 and the corresponding $\Gamma^2 = 0.422$.

A convenient notation is

$$l_c = \frac{v_c}{S_b} \qquad (7\text{-}6a)$$

Note that

$$\frac{dl_c}{dt} = r \qquad (7\text{-}6b)$$

and with the further notation

$$K = \frac{S_b}{A_t} \qquad (7\text{-}6c)$$

The combination of Eq. 7-4, 7-5, and 7-6 yields

$$\frac{dp_o}{dt} = \frac{p_p}{l_c} \left[\left(1 - \frac{p_o}{p_p} \right) r - \frac{\Gamma^2}{K} c^* \frac{p_o}{p_p} \right] \qquad (7\text{-}7)$$

[3] For an ideal propellant of the type considered here, p_p is related to the maximum explosion pressure by: p (maximum explosion) $= \gamma p_p$, where γ is the ratio of the specific heats (constant pressure/constant volume) for the combustion products. The parameter p_p, in some degree, is an index of merit for the propellant. For asphalt potassium perchlorate propellants it is 125,000 lb/in.² abs and for ballistite it is 230,000 lb/in.² abs.

This equation, together with Eq. 7-6b, governs the internal ballistics of the end-burning rocket.

It is most convenient to analyze Eq. 7-7 by considering the two terms in the parentheses separately. Define $\dot{m}_b$, $\dot{m}_o$ as follows:

$$\dot{m}_b = \left(1 - \frac{p_c}{p_p}\right) r \tag{7-8a}$$

$$\dot{m}_o = \frac{\Gamma^2}{K} c^* \frac{p_c}{p_p} \tag{7-8b}$$

The function $\dot{m}_b$ is proportional to the net rate at which mass enters the combustion chamber due to combustion of the solid propellant. The term

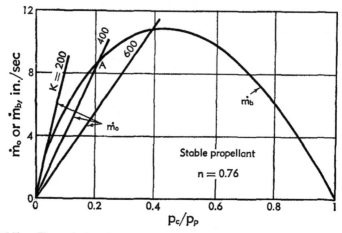

Fig. H,7b. The variation of net mass burning rate m_b, and of the net mass outflow $\dot{m}_o$, with chamber pressure. (Assumed characteristics: $a = 0.937$ in./sec, $n = 0.76$, $c^* = 3750$ ft/sec, $\gamma = 1.2$, $p_p = 125,000$ lb/in.² abs.)

$[1 - (p_c/p_p)] = [1 - (\rho_c/\rho_p)]$ is a correction to the gross burning rate to take into account the volume previously occupied by the solid propellant, which, after combustion, must be filled with combustion gas. The function $\dot{m}_o$ is proportional to the rate at which mass leaves the combustion chamber through the exhaust nozzle. Clearly, if $\dot{m}_o = \dot{m}_b$, then $dp_c/dt = 0$, and the chamber pressure is constant. The question of the stability jo this equilibrium will now be discussed.

In order to represent $\dot{m}_b$, the expression for the burning rate, Eq. 5-3, must be used:

$$r = a \left(\frac{p_c}{1000}\right)^n \tag{5-3}$$

Assume first that $0 < n < 1$. Then the functions $\dot{m}_b$ and $\dot{m}_o$ may be plotted as a function of p_c as shown in Fig. H,7b. The function $\dot{m}_o$ is plotted for three values of the area ratio S_b/A_t. Consider the intersection of the $\dot{m}_o$ and $\dot{m}_b$ curves at A. A disturbance, leading to a slight increase in chamber pressure, results in a situation in which $\dot{m}_o > \dot{m}_b$; the mass outflow exceeds the rate of mass combustion. Therefore (see Eq. 7-7 and 7-8) the chamber pressure decreases until the point A is reached. Similarly, if a disturbance leading to a decrease in pressure occurs, $\dot{m}_o < \dot{m}_b$, and the pressure rises until point A is reached. The steady state condition, $\dot{m}_o = \dot{m}_b$, is stable as long as the shape of the $\dot{m}_b$ curve is as indicated in Fig. H,7b. The shape of the curve is determined by the exponent n in the burning rate expression, Eq. 5-3.

The fundamental requirement for stability is that at the point of intersection of the $\dot{m}_o$ and $\dot{m}_b$ curves, the steady state point, the slope of the $\dot{m}_o$ curve shall be larger than the slope of the $\dot{m}_b$ curve. At the pressure p_o, defined by

$$\frac{\Gamma^2}{K} c^* \frac{p_o}{p_p} = \left(1 - \frac{p_o}{p_p}\right) a \left(\frac{p_o}{1000}\right)^n \tag{7-9a}$$

there must obtain:

$$\frac{d}{dp_o}\left(\frac{\Gamma^2}{K} c^* \frac{p_o}{p_p}\right) > \frac{d}{dp_o}\left[\left(1 - \frac{p_o}{p_p}\right) a \left(\frac{p_o}{1000}\right)^n\right] \tag{7-9b}$$

By carrying out the differentiation and utilizing Eq. 7-9, the inequality may be written as

$$n < \frac{1}{\left(1 - \frac{p_o}{p_p}\right)} \tag{7-10a}$$

The parameter p_p has numerical values ranging from 100,000 to 300,000 lb/in.^{2} abs, so that for all situations of practical interest, $p_o < 3000$ lb/in.^{2} abs, say, the ratio p_o/p_p is a small number. The essential stability requirement is then

$$n < 1 \tag{7-10b}$$

It is true that even if $n > 1$ there will always be one stable steady state pressure. However, this will lie at an unacceptably high value. Taking the equality sign in Eq. 7-10a and solving for the pressure yields

$$p_o \text{ (minimum, stable)} = \frac{n-1}{n} p_p \tag{7-11}$$

If $n < 1$, Eq. 7-11 predicts negative stable pressures. This is merely mathematical continuation, but the physical meaning is that a rocket unit could operate in a stable manner at any required low pressure. If $n = 1$, stable pressures down to zero are predicted. But consider the case $n = 1.1$, $p_p = 125,000$ lb/in.^{2} abs (a moderate value), then the minimum stable pressure is at 11,300 lb/in.^{2} abs, a value too high for practical rocket design.

An example for hypothetical propellant with a burning rate exponent, $n = 2$, is indicated in Fig. H,7c. The intersections of the $\dot{m}_o$, $\dot{m}_b$ curves, marked U, are unstable. A slight increase in pressure results in more rapid mass production than can escape from the exhaust nozzle. The intersection points marked S are stable steady pressure points. The minimum steady stable pressure corresponds to tangency of the $\dot{m}_o$, $\dot{m}_b$ curves and for this example is at $\frac{1}{2}p_p \cong 62{,}500$ lb/in.^{2} abs. This also indicates that there is a minimum area ratio $S_b/A_t = K$ ($= 80$ for this example) below which $\dot{m}_o$ is always greater than $\dot{m}_b$. A rocket designed with $K < 80$

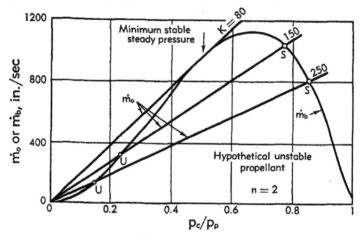

Fig. H,7c. The variation of net mass burning rate $\dot{m}_b$, and of net mass outflow $\dot{m}_o$, with chamber pressure for a hypothetical unstable propellant. (Assumed characteristics: $a = 0.0473$ in./sec, $n = 2$, $c^* = 3750$ ft/sec, $\gamma = 1.2$, $p_p = 125{,}000$ lb/in.^{2} abs.)

would always stop burning immediately after ignition (for this hypothetical propellant).

Experimental propellants have been prepared for which $n > 1$, and the prediction of very high pressure development has been verified. The curves (Fig. H,7c) also indicate difficulty in igniting the rocket since pressure, during the ignition phase, must rise at least to U since $\dot{m}_o > \dot{m}_b$ for lower pressures.

These curves (Fig. H,7c and H,7b) have been computed assuming that the ideal expressions for $\dot{m}_o$, $\dot{m}_b$ remain valid for the whole pressure range up to p_p. This assumption is known to be incorrect at low pressures (see Art. 12 on the combustion limit) and is probably wrong at high pressures, but it does illustrate the whole regime of rocket operation.

The criterion established for n ($n \leq 1$) for the stability, is theoretically correct. However, values of n near unity lead to extreme sensitivity of the rocket to small disturbances. Low values of n are desired, and a propellant for which $n > 0.8$, say, is rather unsuitable.

Variation of chamber pressure with area ratio. Having established the stability criterion, one may now compute the stable steady pressure as a function of K, the ratio of the burning surface area to the exhaust nozzle throat area. The steady pressure obtains when $dp_c/dt = 0$ in Eq. 7-7. The resulting expression may be arranged to be

$$K = \frac{\Gamma^2 c^* \dfrac{p_c}{p_p}}{\left(1 - \dfrac{p_c}{p_p}\right) r} = \frac{p_c}{\rho_p c^* \left(1 - \dfrac{p_c}{p_p}\right) r} \qquad (7\text{-}12)$$

Experimental values of r and c^* as a function of pressure may be used, or Eq. 5-3 and 7-5 may be employed (p_p may be eliminated by utilizing

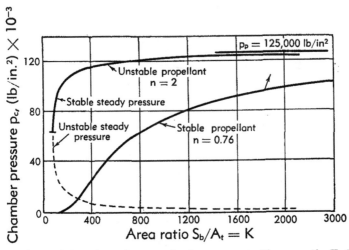

Fig. H,7d. Variation of steady state chamber pressure with area ratio K, for a stable and for an unstable propellant. The unstable propellant exhibits no stable steady state pressure below 62,500 lb/in.² abs. (The assumed propellant characteristics are given in Fig. H,7b and H,7c.)

$p_p = \rho_p \Gamma^2 (c^*)^2$). Numerically, it is easier to assume p_c and compute K. Examples of the area ratio curves corresponding to Eq. 7-12 are shown for a stable and an unstable propellant in Fig. H,7d. The scale employed again covers the whole range of pressures. For practical rocket work, curves not extending above 4000 lb/in.² abs are all that are needed. An illustrative set is shown in Fig. H,7e. The very rapid variation of chamber pressure with changes in K may be noted.

In the low pressure region, $p_c \ll p_p$, the expression for the burning rate (Eq. 5-3) may be inserted into Eq. 7-12 and the expression solved for the chamber pressure as a function of area ratio. As noted above,

p_p may be eliminated by utilizing Eq. 7-3 and 7-5. The result is

$$\frac{p_c}{1000} = \left(\frac{\rho_p a c^* K}{1000}\right)^{\frac{1}{1-n}} \tag{7-13}$$

where a and n are the parameters in the burning rate law. This indicates that for $n \cong 0.75$, the chamber pressure varies with the 4th power of

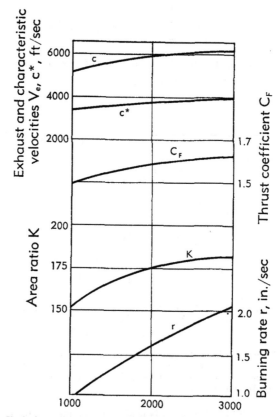

Fig. H,7e. Variation of propellant performance characteristics with combustion chamber pressure for an asphaltic, thermoplastic, composite propellant (end-burning).

the area ratio, giving a very rapid change of pressure with area ratio. This indicates the sensitive balance between gas evolution and outflow in a rocket motor, even for stable propellants. In experimental studies, data have been analyzed by log p_c vs. log K and log r vs. log p_c plots and fitted by least square methods. In a typical example n from K vs. p_c data was 0.80 and n from r vs. p_c data was 0.76. The higher value of n for the

K vs. p_c plot is probably due to the increase of c^* with chamber pressure indicated in Fig. H,7e.

H,8. Steady State Dynamics for Radial-Burning Grains. For internal-burning and radial-burning grains, the internal gas flow introduces complications into the theory. Erosive burning, if it is mass flow dependent, can also introduce a special kind of instability—erosive instability—into the combustion process.

A definite arrangement, of the type considered here, is shown in Fig. H,8a. The analysis extends in a straightforward manner to other, related, arrangements.

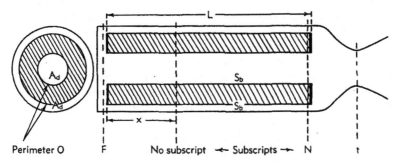

Fig. H,8a. Notation employed in the discussion of radial-burning solid propellant grains.

If the burning rate remains the same at every point on the surface, a hollow cylindrical charge, restricted from burning on the ends, maintains a constant burning surface throughout the burning period. Other arrangements can be devised to achieve something of the same result. It is assumed that the total area of the burning surface S_b is constant.

Even with this assumption, the gas flows do not correspond to steady state conditions. As the solid propellant burns away, the duct area A_d, for gas flow parallel to the axis of the rocket, increases. However, in the usual design regime for radial-burning rockets, the velocity of gas flow at the exhaust nozzle end of the charge is very much larger than the velocity of recession of the burning surface. During the time necessary for the combustion of 1 per cent of the solid propellant, the gas evolution corresponds to about twice the duct volume within the motor. That is, during the time of passage of a fluid particle from the burning surface to the exhaust nozzle throat, the duct area changes by only $\frac{1}{2}$ per cent. This would seem to indicate that assumption of steady state gas flow conditions for interior ballistics analysis is a satisfactory approximation, and steady state analysis is employed here.

In addition, it is assumed that the properties of the combustion gas are the same across any given cross section of the duct. This implies mixing and a high degree of turbulence. These conditions are usually fulfilled for rocket motor gas flows.

Finally, ideal gas behavior is assumed for the propellant gas. Actually, even if the gas composition changes as a function of pressure and temperature, its behavior can be closely approximated by selecting a suitable value for the ratio of specific heats [10].

The skin drag and frictional forces of the internal gas flow on the propellant grain and on the motor tube are neglected. The Reynolds number for internal gas flows in rockets is of the order of 10^6 and the corresponding friction factor is about 0.005. A calculation indicates that, during the critical period immediately after ignition, when the duct area is small, the frictional loss corresponds to 2 to 5 per cent of the pressure drop along the grain due to the dynamics of the gas flow. This omission of frictional effects, then, does introduce appreciable numerical error, but the qualitative features of rocket behavior are not affected.

Because of the internal flow, the analysis is complex. By defining an average value for the burning rate, explicit expressions are easily derived. The more intricate problem of finding suitable expressions for the average burning rate and related functions is then carried out as a separate problem and inserted into the preceding analysis.

Area ratio expressions. The area ratio curve is easily derived by the introduction of the concept of average burning rate and exhaust nozzle stagnation pressure. Consider the arrangement in Fig. H,8a. Between the front of the combustion chamber, marked F, and the exhaust nozzle end of the charge, marked N, the pressure of the gas stream decreases and the velocity of the gas flow increases. The actual distribution will be derived, shortly, but from the section N to the section t at the exhaust nozzle throat no further combustion mass is added to the flow. The flow between N and t corresponds to flow through a portion of an exhaust nozzle (of rather peculiar shape it is true) in which the entering gas velocity is not zero. The sudden expansion of gas as it leaves the end of the grain and enters the region just forward of the exhaust nozzle actually results in loss of stagnation pressure, but this is neglected in the analysis. Now the mass flow through an exhaust nozzle is given by

$$\dot{m}_o = \frac{p^0 A_t}{c^*} \tag{8-1}$$

The stagnation pressure p^0 is related to the pressure and velocity at section N, and through this to conditions at section F. The relation will be derived shortly.

The mass flow out of the exhaust nozzle must equal the rate of mass combustion, which in terms of the average of the burning rate over the propellant surface is given by

$$\dot{m}_b = \bar{r} S_b \rho_b \tag{8-2}$$

Here $\bar{r}$ is the average burning rate, S_b the area of the burning surface, and ρ_p the density of the solid propellant. Eq. 8-2 neglects the effect of increase of duct volume as the charge burns away. This would introduce a factor $[1 - (\rho/\rho_p)]$ on the right (ρ is the density of gas in the ducts), but at normal design pressures $\rho/\rho_p < 0.01$, and this factor is neglected. Equating mass outflow to mass burned gives

$$\frac{p^0 A_t}{c^*} = \bar{r} S_b \rho_p$$

This may be written, in terms of $K = S_b/A_t$,

$$K = \frac{p^0}{\rho_p \bar{r} c^*} \tag{8-3}$$

It is now convenient to introduce the parameters θ_r and θ^*, defined by

$$\theta^* = \frac{p_F}{p^0} \quad \text{or} \quad p^0 = \frac{p_F}{\theta^*} \tag{8-4a}$$

$$\theta_r = \frac{\bar{r}}{r_F} \quad \text{or} \quad \bar{r} = \theta_r r_F \tag{8-4b}$$

$$\theta = \theta^* \theta_r \tag{8-4c}$$

Here r_F and p_F represent the burning rate and gas pressure at the front of the grain, section F (see Fig. H,8a). In terms of these parameters,

$$K = \frac{p_F}{\rho_p c^* \theta^* r_F \theta_r} = \frac{p_F}{\rho_p c^* r_F \theta} \tag{8-5}$$

The parameters θ_r, θ^* may be determined in terms of the ratio to gas duct area to exhaust nozzle throat area A_d/A_t, and the front chamber pressure. Therefore Eq. 8-5 represents the area ratio curve for front pressure.

Internal gas flow. In general, the burning rate is not the same at every section of the grain, so that as burning progresses the gas duct area is different from section to section. This variation is neglected. However, the conditions at the nozzle end of the grain are correctly given although averages taken over the length of the grain are slightly in error. That is, the parameter θ^* is correct, but θ_r and θ are slightly in error due to this approximation. These errors are usually small, relative to the errors introduced by the neglect of friction effects.

The analytical technique used is to find the variation in conditions along the grain, relative to front conditions, in terms of the local Mach number. The Mach number at the nozzle end of the grain is determined by the ratio of exhaust nozzle throat area to the gas duct area. This then unifies the entire flow pattern.

The equations for conservation of mass, momentum, and energy at section x (Fig. H,8a), together with the equation of state, are, respectively

$$A_d \frac{d(\rho V)}{dx} = \rho_b r \frac{dS_b}{dx} = \rho_b r O \tag{8-6a}$$

$$OL = S_b \tag{8-6b}$$

$$p_F - p = \rho V^2 \tag{8-6c}$$

$$\tfrac{1}{2} V^2 = c_p (T_0 - T) \tag{8-6d}$$

$$p = \rho \Re T \tag{8-6e}$$

Here ρ, V, p, and T are the gas density, velocity, pressure, and temperature at section x. A_d is the gas duct area, r the burning rate at position x, dS_b/dx is constant, and O the perimeter of the burning surface. $\Re$ is the engineering gas constant (includes the effect of molecular weight), $T_0 (= T_F)$ is the adiabatic constant pressure flame temperature for combustion of the solid propellant, the subscript F refers to front conditions, and the other symbols have been defined previously.

Eq. 8-6b is exact if the gas duct area is constant along the grain. In general the grain tapers slightly during combustion, producing an opposite taper in the gas duct.[4] This effect is neglected.

It is convenient to introduce the local Mach number M and the dimensionless mass flow density g, defined as follows:

$$M^2 = \frac{V^2}{a^2} = \frac{V^2}{\gamma R T} \tag{8-7a}$$

$$g = \frac{\rho V}{\rho_F a_F} \tag{8-7b}$$

Here a is the local velocity of sound in the gas (not to be confused with the burning rate parameter), γ is the ratio of specific heats (constant pressure/constant volume) for the propellant gas, a_F is the velocity of sound in the gas at section F, and ρ_F is the gas density at section F.

With these parameters Eq. 8-6 can all be solved for local conditions in terms of front conditions and local Mach number. The parameter g can also be expressed in terms of local Mach number, so that the variation of condition along the grain is now determined as a function of g.

---

[4] The burning rate is affected by compensating factors. At the front, the higher pressure tends to accelerate the rate; and at the rear, lower pressure region, the erosive effect of gas flow accelerates the rate. Usually the compensation is not exact.

The results are[6]:

$$\frac{p}{p_F} = \frac{1}{1 + \gamma M^2} \tag{8-8a}$$

$$\frac{T}{T_F} = \frac{T}{T_0} = \frac{1}{1 + \dfrac{\gamma - 1}{2} M^2} \tag{8-8b}$$

$$\frac{\rho}{\rho_F} = \frac{1 + \dfrac{\gamma - 1}{2} M^2}{1 + \gamma M^2} \tag{8-8c}$$

$$\frac{V}{a_F} = \frac{M}{\sqrt{1 + \dfrac{\gamma - 1}{2} M^2}} \tag{8-8d}$$

$$g = \frac{M}{1 + \gamma M^2} \sqrt{1 + \frac{\gamma - 1}{2} M^2} \tag{8-8e}$$

$$M^2 = \frac{\sqrt{1 - 2(\gamma + 1)g^2} - (1 - 2\gamma g^2)}{[(\gamma - 1) - 2\gamma^2 g^2]} \tag{8-8f}$$

The local stagnation pressure p^0 is defined by

$$\frac{p^0}{p} = \left(\frac{T_F}{T}\right)^{\frac{\gamma}{\gamma - 1}}$$

and may be expressed in terms of p_F to give

$$\frac{p^0}{p_F} = \frac{\left(1 + \dfrac{\gamma - 1}{2} M^2\right)^{\frac{\gamma}{\gamma - 1}}}{1 + \gamma M^2} \tag{8-8g}$$

The local stagnation pressure lies about halfway between the local pressure and the front pressure. If $M \ll 1$, one may expand (Eq. 8-8g) and dropping higher orders of M^2 find

$$p^0 = p_F - \tfrac{1}{2}(p_F - p)$$

In the limiting case, $M = 1$, the factor $\tfrac{1}{2}$ is replaced by 0.36. The approximate validity of the factor $\tfrac{1}{2}$ is reasonable since mass is being added along the gas streams at varying pressures above p, and a linear average would give just this value.

In Fig. H,8b the flow characteristics are plotted as a function of local Mach number. The local Mach number and other flow parameters are shown in Fig. H,8c plotted as functions of the dimensionless mass flow density g.

Internal sonic flow. The range of Mach numbers shown does not

[6] The algebra is simplied if expressions for p and T in terms of M are found first. See [44].

exceed unity. Physically, supersonic flow is not attained adjacent to a burning surface where mass is being added. Differentiation of Eq. 8-8e yields

$$\frac{dM}{M} = \frac{(1 + \gamma M^2)\left(1 + \frac{\gamma - 1}{2} M^2\right)}{1 - M^2} \frac{dg}{g} \tag{8-9}$$

It is clear that if $M < 1$, the Mach number increases with g; but if $M > 1$, the Mach number decreases as the mass flow increases. The flow cannot

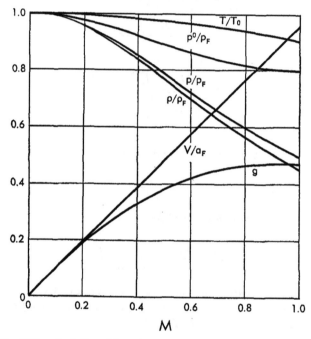

Fig. H,8b.   Variation of flow characteristics with local Mach number.

become supersonic. At $M = 1$, the critical value of g is $\sqrt{1/2(\gamma + 1)}$. If the burning surface or the burning rate is large enough, or if the duct area is small enough, the conditions for sonic flow may be met for some point along the grain. The sonic position moves to the nozzle end of the grain and then jumps to the exhaust nozzle throat, provided that $A_t < A_d$. If $A_t > A_d$, then a sonic flow will be established near the nozzle end of the grain. Variation of the duct area, due to variation in burning rate, may cause the sonic position to jump from place to place along the grain. Uniform flow can be established, if $A_d > A_t$ and if the pressure level rises, since g is dimensionless and the actual mass flow can increase with increasing pressure. Since sonic flow corresponds to $M = J = 1$ (see below), the expression for the front pressure becomes

$$\frac{S_b}{A_d} = \frac{p_F}{\rho_p c^* r_F \theta}$$

where $\theta = \theta^* \theta_r = \theta^*(1)\theta_r(1, p_F)$. The value 1 corresponds to $J = 1$. This would enable one, in principle, to solve for the steady state front chamber pressure. The usual area ratio K is replaced by S_b/A_d.

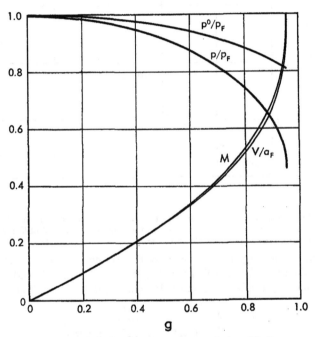

Fig. H,8c. Variation of flow characteristics with the dimensionless mass flow density parameter g.

It is clear that the internal geometry of the rocket should be arranged so that supersonic flow is not attained at any section of the grain. This is accomplished by insuring that $A_d > A_t$ at all times. Difficulties, such as unstable combustion and loss of propellant, may be encountered if this requirement is not met.

Throat-to-duct area ratio. Having established that sonic flow will occur only at the exhaust nozzle throat, it is easy to determine the Mach number M_N at the exhaust nozzle end of the propellant. The equality of mass flow at section N and at the nozzle throat, section t, yields (using Eq. 8-1)

$$\rho_N V_N A_d = \dot{m}_o = \frac{p_N^0 A_t}{c^*}$$

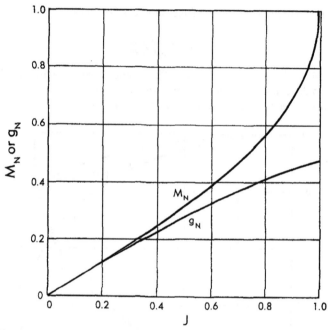

Fig. H,8d. Variation of Mach number at the exhaust nozzle end of the charge, and of the dimensionless mass flow parameter at the exhaust nozzle end of the charge, g_N with the ratio, J of exhaust nozzle throat area to gas duct area (for $\gamma = 1.2$).

Here p_N^0 is the stagnation pressure of the gas flow at the exhaust nozzle end of the grain. By utilizing the definition of c^* (Eq. 7-5) and Eq. 8-8c, 8-8d, and 8-8g the Mach number M_N may be determined from

$$J = \frac{A_t}{A_d} = \frac{M_N}{\left(\dfrac{2}{\gamma + 1} + \dfrac{\gamma - 1}{\gamma + 1} M_N^2\right)^{\frac{\gamma+1}{2(\gamma-1)}}} \tag{8-10}$$

From the definition (Eq. 8-4), θ^* may be written (Eq. 8-8g)

$$\theta^* = \frac{1 + \gamma M_N^2}{\left(1 + \dfrac{\gamma - 1}{2} M_N^2\right)^{\frac{\gamma}{\gamma-1}}} = \theta^*(J) \tag{8-11}$$

By knowing M_N from Eq. 8-10, θ^* may be expressed as a function of the throat-to-duct area ratio as indicated.

The variation of M_N with J is shown in Fig. H,8d, and the variation of θ^* with M_N and with J is shown in Fig. H,8e.

The average burning rate and $g(x)$. If the burning rate were the same at all sections along the grain, then the actual mass flow density ρV and the dimensionless mass flow density g would be proportional to x, the distance from the front of the grain (Fig. H,8a). The parameter g is the natural independent variable for description of the internal gas flow and local burning rate (g and M are connected and either may be used. See Eq. 8-8).

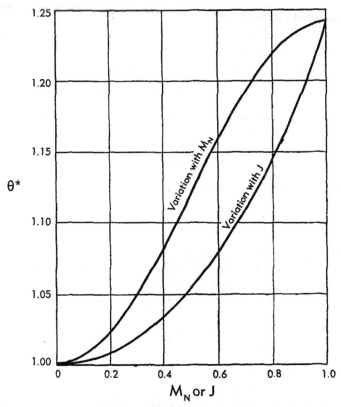

Fig. H,8e.   Variation of the stagnation pressure flow parameter θ^* with Mach number at the exhaust nozzle end of the charge and also with the ratio J of exhaust nozzle throat area to gas duct area (for $\gamma = 1.2$).

The connection between g and x is derived first. The burning rate expressions (Eq. 5-1 and 5-2) may be written in terms of the front pressure and g. Let this be understood, then Eq. 8-6a may be written

$$dx = \frac{A_d}{\rho_p O} \frac{d(\rho V)}{r}$$

From the definition of g (Eq. 8-7b),

$$dx = \frac{A_d \rho_F a_F}{\rho_p O} \frac{dg}{r}$$

By Eq. 8-6b, and integration, this becomes

$$\frac{x}{L} = \frac{A_d}{S_b} \frac{\rho r a_F}{\rho_p} \int_0^g \frac{dg}{r} \tag{8-12}$$

This is the connection between the mass flow parameter and x. It may be simplified by calculation of the average burning rate $\bar{r}$.

The rate of mass flow past the nozzle end of the grain must equal the rate at which mass is consumed over the entire burning surface. Introducing the average burning rate $\bar{r}$ and the definition of g, this gives

$$\bar{r}\rho_p S_b = A_d(\rho V)_N = A_d \rho r_F a_F g_N \tag{8-13}$$

If the integration in Eq. 8-12 is extended to g_N, then $x = L$. Combining Eq. 8-13 and 8-12 then gives

$$\frac{1}{\bar{r}} = \frac{1}{g_N} \int_0^{g_N} \frac{dg}{r} \tag{8-14a}$$

From the definition of θr (Eq. 8-4b),

$$\frac{1}{\theta_r} = \frac{1}{g_N} \int_0^{g_N} \frac{dg}{(r/r_F)} \tag{8-14b}$$

Now r/r_F depends only on M (or g), so that the integration may be carried out explicitly (although numerically). The burning rate expressions employed here may be combined with the flow functions (Eq. 5-1, 5-2 and 8-8) to give, for mass flow density erosion,

$$\left(\frac{r}{r_F}\right)_{\mathrm{me}} = \left(\frac{1}{1+\gamma M^2}\right)^n (1 + k\rho r_F a_F g) \tag{8-15a}$$

and for velocity erosion,

$$\left(\frac{r}{r_F}\right)_{\mathrm{ve}} = \left(\frac{1}{1+\gamma M^2}\right)^n f_3\left(a_F \frac{V}{a_F}\right) \tag{8-15b}$$

By substituting the expressions for a_F, ρ_p in terms of c^*, p_p (Eq. 7-3 and 7-5), and by employing the expression for V/a_F (Eq. 8-8d), the burning rate ratios may be written, for mass flow density erosion,

$$\left(\frac{r}{r_F}\right)_{\mathrm{me}} = \left(\frac{1}{1+\gamma M^2}\right)^n \left(1 + \lambda_1 \frac{p_F}{p_p} g\right) \tag{8-16a}$$

$$\lambda_1 = \sqrt{\gamma}\, \Gamma k \rho_p c^* \tag{8-16b}$$

so that ultimately

$$(\theta_r)_{\mathrm{me}} = (\theta_r)_{\mathrm{me}}\left(g_N, \frac{p_F}{p_p}\right) = (\theta_r)_{\mathrm{me}}\left(J, \frac{p_F}{p_p}\right) \tag{8-16c}$$

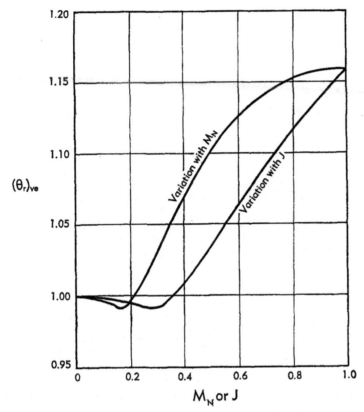

$(\theta_r)_{ve}$

M_N or J

Fig. H,8f. Velocity erosion. Variation of the velocity erosive burning rate parameter θ_r with Mach number at the exhaust nozzle end of the charge and also with the ratio J of exhaust nozzle throat area to gas duct area. (For $\gamma = 1.2$, $a = 0.651$ in./sec, $n = 0.69$, $c^* = 5200$ ft/sec, $V_0 = 600$ ft/sec, $k_v = 5.72 \times 10^{-4}$ sec/ft, $\rho_p = 101.5$ lb/ft³, $p_p = 230{,}000$ lb/in.² abs.)

and for velocity dependent erosion,

$$\left(\frac{r}{r_F}\right)_{ve} = \left(\frac{1}{1 + \gamma M^2}\right)^n f_3(u) \qquad (8\text{-}17a)$$

$$f_3(u) = \begin{cases} 1 & , u < u_0 \\ 1 + \lambda_2(u - u_0), & u > u_0 \end{cases} \qquad (8\text{-}17b)$$

$$u = \frac{M}{\sqrt{1 + \frac{\gamma - 1}{2} M^2}} \qquad (8\text{-}17c)$$

$$\lambda_2 = \sqrt{\gamma}\,\Gamma k_v c^* \qquad (8\text{-}17d)$$

$$u_0 = \frac{V_0}{\sqrt{\gamma}\,\Gamma c^*} \qquad (8\text{-}17e)$$

so that ultimately

$$(\theta_r)_{ve} = (\theta_r)_{ve}(g_N) = (\theta_r)_{ve}(J) \tag{8-17f}$$

The new erosion constants λ_1 and λ_2 are now dimensionless. The burning rate factor may be computed by combining Eq. 8-12, 8-16, and 8-17. The results of this computation are shown in Fig. H,8f for a velocity-dependent erosive propellant; and in Fig. H,8g for a mass flow density erosive propellant. The former does not depend upon the pressure, but the latter shows a pressure dependence (Eq. 8-16).

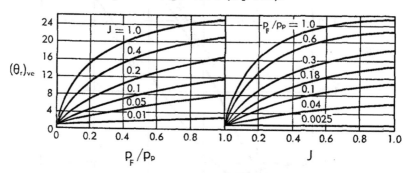

Fig. H,8g. Mass flow density erosion. Variation of the mass flow density erosive burning rate parameter θ_r with chamber pressure at the forward end of the charge p_F, and also with the ratio J of exhaust nozzle throat area to gas duct area. (For $\gamma = 1.2$, $a = 0.15$ in./sec, $n = 0.50$, $c^* = 3650$ ft/sec, $k = 0.61$ in.² sec/lb, $\rho_p = 103$ lb/ft³, $p_p = 112{,}000$ lb/in.² abs.)

The expression (Eq. 8-12) connecting x and g may be simplified by dividing through by the same equation for $x = L$ and $g = g_N$. This gives

$$\frac{x}{L} = \frac{\displaystyle\int_0^g \frac{dg}{r}}{\displaystyle\int_0^{g_N} \frac{dg}{r}}$$

Multiplying numerator and denominator by r_F, and letting $\theta_r(g)$ correspond to the integral (Eq. 8-14b) for $g = g$ rather than $g = g_N$ gives

$$\frac{x}{L} \frac{\theta_r(g_N)}{\theta_r(g)} \frac{g}{g_N} \tag{8-18}$$

This gives x/L in terms of the integral, which would be calculated anyway as a function of g_N for various values.

It is interesting to note that x/L is nearly proportional to g/g_N, differing only by a factor equal to the mean burning rate divided by the mean of the burning rate from the front up to the section at which the mass flow is g. In Fig. H,8h the variation of g with x is shown for a mass flow density erosive propellant.

Comments on the average burning rate dependence. The flow pattern in the rocket is essentially kinematic and geometrical. The explosion enthalpy $c_p T_0 = c_p T_F$ of the propellant determines the speed of sound a_F at the forward end of the grain. This normalizes the velocity distribution and also the connection between density and pressure. With this established, and the pressure level set by the front pressure, all other parameters depend only on local Mach number M or the dimensionless mass flow density g. These are uniquely connected and are fixed by the geometrical ratio of throat-to-duct area ratio. For propellants subject to

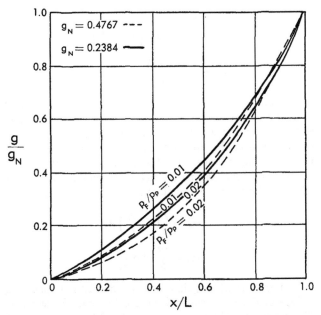

Fig. H,8h. Variation of the dimensionless mass flow density parameter g with the geometric position x, measured from the front of the grain for a mass flow density erosive propellant. (For propellant properties listed in Fig. H,8g.)

velocity erosion alone, the over-all effect is to simply modify the constants for end-burning grains. For propellants subject to mass flow density erosion, the density introduces an additional pressure dependence on the burning rate. The remaining effects are kinematic and geometric.

H,9. Area Ratio Dependence. Erosive Instability. In the case of unrestricted burning rockets, the erosive effect of internal gas flow is encountered. If the erosive effect depends only on gas velocity, the stability considerations are the same as for end-burning grains (θ_r depends only on J) and no new effects are encountered. However, if the erosive effect depends upon the mass flow density ρV (Eq. 3-2b), the over-all effect is to introduce additional pressure dependence for the average burning rate. The additional pressure dependence enters through the density factor of the erosive term. The erosive instability is a more fundamental limitation of rocket design than resonant burning since it is connected with a first order differential equation, and is based upon the conservation of mass.

In order to discuss the whole pressure regime, it is convenient to reintroduce the factor $1 - (p/p_b)$ which takes account of the increase of gas volume within the combustion chamber as the charge burns away. The area ratio curve may be expressed as

$$K = \frac{p_F}{p_b c^* \theta^* r_F \theta_r \left(1 - \dfrac{p_F}{p_b}\right)} \tag{9-1a}$$

$$\theta_r = \theta_r \left(J, \frac{p_F}{p_b}\right) \tag{9-1b}$$

Here p_F is the gas pressure at the forward end of the charge. J is the ratio A_t/A_d of nozzle throat area to gas duct area along the charge (see Fig. H,8a), and θ_r is determined numerically (Fig. H,8g).

This equation is of the same form as Eq. 7-12 for end-burning grains and has the same significance. The stability problem may be analyzed in the same way. The expressions for net mass production $\dot{m}_b$, and mass outflow $\dot{m}_o$, may be written by analogy:

$$\dot{m}_b = \left(1 - \frac{p}{p_b}\right) \theta_r r_F \tag{9-2a}$$

$$\dot{m}_o = \frac{\Gamma^2}{K} \theta^* c^* \frac{p}{p_b} \tag{9-2b}$$

The stability requirement is again that at the steady state point, $\dot{m}_o = \dot{m}_b$, that the slope of the $\dot{m}_o$ curve shall be greater than the slope of the $\dot{m}_b$ curve. The procedure is identical to Eq. 7-9. Although numerical results are available, it is helpful to obtain an approximate expression, analogous to Eq. 7-10. From Eq. 8-14 and 8-16 the expression for θ_r is

$$\frac{1}{\theta_r} = \frac{1}{g_N} \int_0^{g_N} \frac{(1 + \gamma M^2)^n dg}{\left(1 + \lambda_1 \dfrac{p_F}{p_b} g\right)}$$

If the factor depending explicitly on Mach number is assumed constant at a suitable average value, this may be evaluated to give

$$\frac{1}{\theta_r} \cong \overline{(1 + \gamma M^2)^n} \frac{\ln (1 + x)}{x} \tag{9-3a}$$

$$x = \lambda_1 g_N \frac{p_F}{p_b} \tag{9-3b}$$

For the stability criteria, the logarithmic derivative of θ_r is required so that the Mach number factor approximately cancels out, as it does exactly to the assumed approximation. This gives

$$\frac{d \ln \theta_r}{d \ln p_F} = -\left[1 - \frac{x}{(1+x) \ln (1+x)}\right] \qquad (9\text{-}4a)$$

$$x = \lambda_1 g_N \frac{p_F}{p_p} \qquad (9\text{-}4b)$$

Taking the logarithmic derivatives of the two mass flow equations (Eq. 9-2), and employing the result (Eq. 9-4) gives the requirement for stability:

$$n - \frac{p_F/p_p}{(1 - p_F/p_p)} + \left[1 - \frac{x}{(1+x) \ln (1+x)}\right] < 1 \qquad (9\text{-}5a)$$

$$x = \lambda_1 g_N \frac{p_F}{p_p} \qquad (9\text{-}5b)$$

The term in the square brackets increases monotonically with x. It is shown in Fig. H,9a. The lowest pressure for which instability will

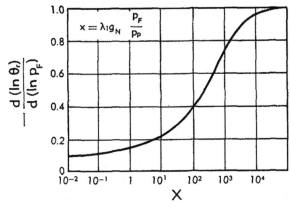

Fig. H,9a. Mass flow density erosive instability. Variation of the correction to the burning rate power index due to mass flow density erosion with dimensionless pressure parameter x.

occur is for the maximum value of g_N (= 0.4767). Setting the left side equal to unity and inserting the propellant characteristics given in Fig. H,8g, the minimum unstable pressure is found at $p_F/p_p = 7.8 \times 10^{-3}$, More accurate calculations employed for Fig. H,8g give 7.4×10^{-3}, which corresponds to a pressure of only 880 lb/in.² abs.

According to Eq. 9-1 the pressure depends upon both K and J. Therefore certain regions of the K, J plane are unstable.

The situation is made clearer if the mass flow functions (Eq. 9-2) are plotted. The result is shown in Fig. H,9b. The erosive effect changes the

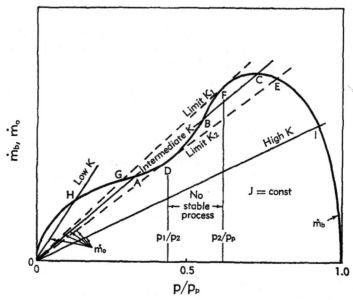

Fig. H,9b. Mass flow density erosive instability. Variation of net burning rate $\dot{m}_b$ and net outflow $\dot{m}_o$ with front chamber pressure p_F (schematic).

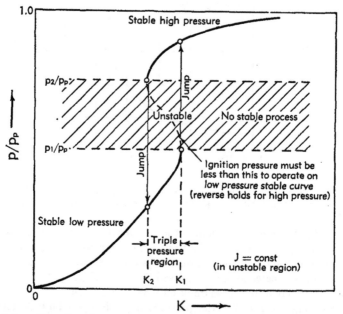

Fig. H,9c. Mass flow density erosive instability. Variation of steady state front pressure p_F with ratio K of burning surface area to exhaust nozzle throat area for a fixed J in the instability region (schematic). Note limit K_1, K_2 and limit p_1, p_2 reversal of magnitude.

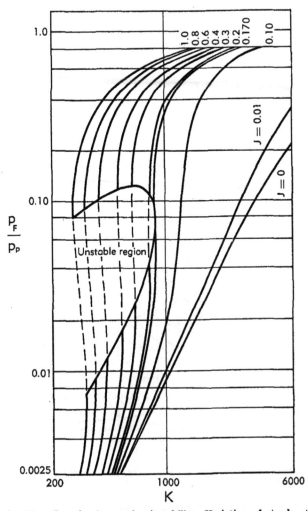

Fig. H,9d. Mass flow density erosive instability. Variation of steady state front pressure with the area ratios K and J. The critical J value is 0.170. Below this value, instability effects are absent. (Propellant characteristics listed in Fig. H,8g.)

$\dot{m}_b$ curve from a simple curve, concave to the pressure axis, to one of more complicated form. If the effect of erosion is changed by changing the ratio of nozzle throat area to gas duct area along the charge, then a *family* of $\dot{m}_b$ curves is developed. For large duct areas, the erosive effect decreases and, below a critical throat-duct area ratio J, the $\dot{m}_b$ curves become simple in character and the erosive instability is no longer encountered.

Now consider the curve shown (Fig. H,9b), which shows a marked erosive effect. The $\dot{m}_o$ and $\dot{m}_b$ intersections at very high and very low K value are simple and stable (intersections H and I). The $\dot{m}_o$ line at intermediate K values makes three intersections, A, B, and C with the $\dot{m}_b$ curve. There are two stable operating pressures A and C, separated by one

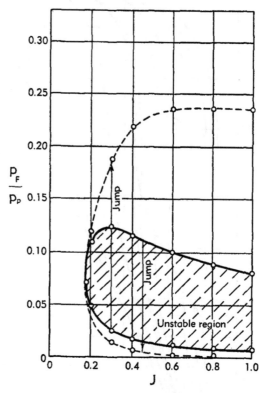

Fig. H,9e. Mass flow density erosive instability. The instability region in the front pressure J plane. The value of K is not fixed, but is adjusted to give the desired pressure. The pressure jump from the low pressure stable curve to the high pressure stable curve, for an infinitesimal change in K at the upper stability limit, is indicated by the arrows and dotted curves. The converse is indicated on going from high to low pressure stable burning. (Propellant properties listed in Fig. H,8g.)

unstable steady pressure B for a rocket of given geometry and given propellant.

If the ignition process is very severe, the ignition pressure may start the operation of the rocket at a pressure in excess of B, and the pressure rises to the stable pressure C, which may be disastrously high. If the ignition pressure lies below pressure B, the rocket drops in pressure and operates at pressure A. The stable pressures fall into two regions, high pressure and low pressure. There are two limiting $\dot{m}_o$ curves just tangent to the $\dot{m}_b$ curve, which are determined by taking the equality sign in Eq. 9-5. It is clear that the point D corresponds to the maximum stable pressure in the low pressure region. If an attempt is made to increase the pressure above D by increasing K (rotating the $\dot{m}_o$ line clockwise), the stable pressure jumps from D to E. (A similar argument follows for points F

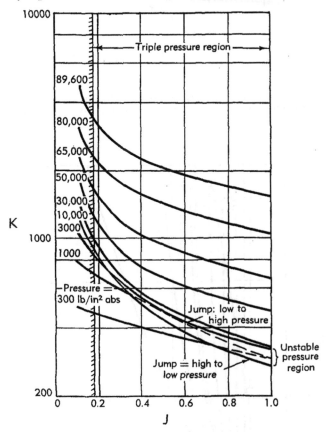

Fig. H,9f. Mass flow density erosive instability. The unstable pressure region in the K, J plane. At the edges of this region, an infinitesimal increase in K causes a discontinuous jump from the stable low pressure region to the stable high pressure region. The converse holds for attempts to continue the high pressure region to low pressures. (Propellant properties listed in Fig. H,8g.)

and G if one attempts to extend the stable high pressure region to low pressures by decreasing K.)

Consider Fig. H,9c, where the variation of steady pressure with K is shown for fixed throat-to-duct ratio. Starting in the low pressure region, the pressure increases with K. A limit K is reached, and the low pressure curve continues smoothly, but now a stable high pressure curve also

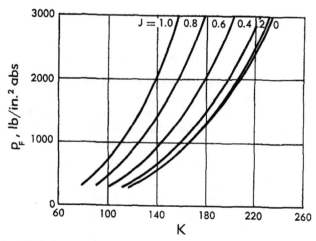

Fig. H,9g. Variation of front chamber pressure with area ratios K and J, for a propellant subject only to velocity erosion. (For propellant properties listed in Fig. H,8f.)

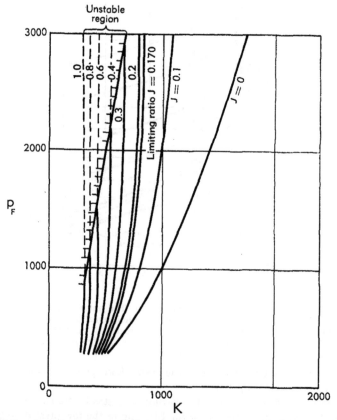

Fig. H,9h. Variation of front chamber pressure with area ratios K and J in the lower stable region, for a propellant subject to mass flow density erosion. (For propellant properties listed in Fig. H,8g.)

appears. As the second limit K is reached, the pressure must jump discontinuously to the high pressure curve. A pressure region is also indicated which cannot be reached by a stable combustion process. To operate on the low pressure curve in the triple (two stable, one unstable) pressure region, the ignition pressure must be less than the unstable pressure. The high pressure curve is of no practical interest because it lies at pressures

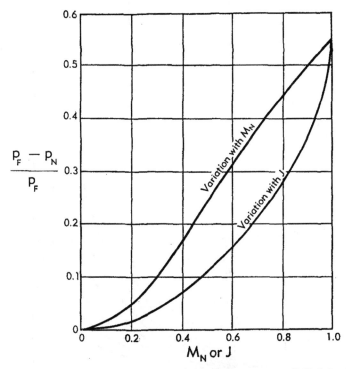

Fig. H,9i. Variation of the ratio: front minus nozzle pressure divided by front pressure with throat-to-duct area J or with Mach number at the nozzle end of the grain, M_N. (For $\gamma = 1.2$.)

comparable with p_p, too high for practical rocket design. The complete family of curves is shown to scale in Fig. H,9d.

In Fig. H,9e the limits of the stable low and high pressure regions are shown as a function of nozzle throat-to-gas duct area ratio. The dotted curves indicate the pressure jump from the stability curve to the high pressure (low pressure) region for a rocket of fixed J and infinitesimal change in K. The critical J value for erosive instability is indicated, for lower J values no erosive instability effects are encountered.

In Fig. H,9f the triple pressure region is outlined, as well as constant pressure contours, on the J, K plane. The critical J value is again indicated, and for J values less than this the $\dot{m}_b$ curve (Fig. H,9b) again has a simple shape.

It is clear that the effect of erosive instability will aggravate the effects of charge deformation and places severe limits on rocket motor design.

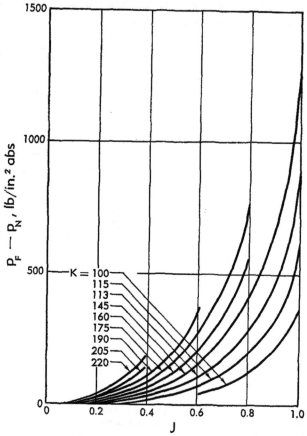

Fig. H,9j. Variation of the pressure difference, front minus nozzle pressure, with throat-to-duct ratio J for various K values, for a velocity-dependent erosive propellant. This is used in the discussion of charge deformation effects. (Propellant properties listed in Fig. H,8f.)

Not only is this a limit on design geometry, but in the triple pressure region it places a limit on ignition pressure.

The limitations implied by the erosive instability phenomenon are somewhat eased in practical rocket operation because the erosive burning enlarges the gas duct area rapidly, particularly at the exhaust nozzle end of the charge. Nevertheless, many blowups can be traced to this effect, charge deformation effects, or a combination of the two.

After the instability characteristics have been studied, the front chamber pressure in the practical low pressure region may be plotted as a function of K and J from Eq. 9-1. The results are shown in Fig. H,9g for a propellant affected only by velocity erosion and therefore not subject to erosive instability. A similar set of curves is shown in Fig. H,9h for a propellant subject to mass flow erosion and erosive instability. After the front pressure is determined, the general variation in pressure may be determined from Eq. 8-8. The front-nozzle pressure difference relative to the front pressure is shown in Fig. H,9i. The absolute front-nozzle pressure difference is shown in Fig. H,9j as a function of J for various values of K for a velocity-dependent erosive propellant. Curves of this type are employed in the discussion of charge deformation problems.

H,10. Temperature Sensitivity, Transients, Thin Web Grains, Resonant Burning, Chuffing, and Gas Generation.

Temperature sensitivity. A rocket of fixed geometry and given propellant shows a variation in performance with changing propellant temperature. This is due to the variation of burning rate with propellant temperature. Increasing the temperature increases the burning rate which increases the pressure, this in turn increases the burning rate further, etc. The essential result may be obtained as follows: At the pressures usual in rocket design, the expression for the area ratio curve may be simplified to yield a power law for the connection between pressure and area ratio. Eq. 7-13 provides this solution:

$$\frac{p_0}{1000} = \left(\frac{\rho_p a c^*}{1000} K\right)^{\frac{1}{1-n}} \tag{7-13}$$

The assumption has been made that the burning rate depends on the propellant temperature T_p only through a. (The indication is that the index n and the erosive constant k are not temperature-dependent for many propellants). Consequently the intrinsic temperature dependence of the burning rate is defined to be

$$\pi_r = \frac{1}{r}\left(\frac{dr}{dT_p}\right)_p = \frac{1}{a}\frac{da}{dT_p} = \frac{1}{T_1 - T_p} \tag{10-1}$$

The subscript p indicates the derivative at constant pressure. The equation also defines the burning rate temperature sensitivity coefficient π_r.

Logarithmic differentiation of Eq. 7-13 gives

$$\pi_p = \frac{1}{p_0}\left(\frac{dp_0}{dT_p}\right)_K = \frac{1}{1-n}\frac{1}{a}\frac{da}{dT_p} \tag{10-2}$$

The subscript K indicates the derivative at constant area ratio. The equation also defines the pressure temperature sensitivity coefficient π_p.

From Eq. 10-1 and 10-2 the result follows

$$\frac{1}{p_o}\left(\frac{dp_o}{dT_p}\right)_K = \frac{1}{1-n}\frac{1}{r}\left(\frac{dr}{dT_p}\right)_p \tag{10-3}$$

The factor $1/(1-n)$ is an amplification factor for the effect of burning rate variation on rocket motor pressure. For real propellants the characteristic velocity c^* increases with combustion pressure; this effect makes the amplification larger. For radial-burning grains, and velocity erosion, the result of Eq. 10-3 is not changed. For mass flow density erosive propellants, however, there is an additional pressure dependence not included in Eq. 7-13. This may be taken into account by replacing n in Eq. 10-3 by the expression which follows from the stability study (Eq. 9-5).

$$n \to n' = n + \left[1 - \frac{x}{(1+x)\ln(1+x)}\right] \tag{10-4a}$$

$$x = \lambda_1 g_N \frac{p_F}{p_p} \tag{10-4b}$$

Geckler and Sprenger [44] have given a critical review of this and related problems.

Start-up and pressure recovery. In the treatment of steady state dynamics, it was assumed that the temperature of the gas in the combustion chamber or at the forward end of the grain did not vary with time. Actually, just before ignition, the combustion chamber is filled with a cold gas, usually air, and for a short time after ignition the gas temperature remains below the flame temperature. An analysis of this complex problem will not be attempted here. Von Kármán and Malina [45] have studied the problem in some detail and find that the average gas temperature rises quite rapidly to its normal value. This is illustrated in Fig. H,10. The time scale is longer the larger the ratio of initial volume to burning surface area.

The important conclusion that may be drawn from Fig. H,10 is that the temperature attains its steady value much more quickly than does the pressure. In further studies it will be assumed that the temperature is always at the adiabatic flame temperature value, which will only slightly distort results at the beginning of combustion.

With this assumption, analysis of pressure variation is simple. From Eq. 7-7, the theoretical value for $p_p = \Gamma^2 \rho_p c^{*2}$, and neglecting the term p_o/p_p compared to unity, follows:

$$\frac{dp}{dt} = \frac{\Gamma^2 \rho_p c^{*2}}{l_o}\left[r - \frac{c^*}{K}\frac{p}{\rho_p c^{*2}}\right] \tag{10-5}$$

The subscript has been omitted from p, which represents the chamber pressure for end-burning grains or the front pressure for radial-burning grains. The equilibrium or steady state condition is

$$K = \frac{p_e}{\rho_p c^* r_e}$$

(10-6)

where the subscript $_e$ designates the steady state value.

The effective burning rate may be written

$$r = a' \left(\frac{p}{1000}\right)^{n'}$$

(10-7)

If no erosive burning is present, set $a' = a$, $n' = n$; if velocity erosive

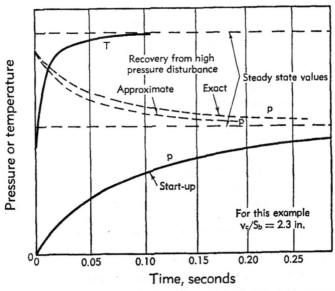

Fig. H,10. Variation of average gas temperature and chamber pressure with time subsequent to ignition (solid curves). Variation of pressure with time subsequent to a high pressure disturbance (constant gas temperature, dashed curves). The "exact" curve takes into account the variation of combustion chamber gas volume with time (dashed curves).

burning and pressure drop along the tube is present, set $a' = a\theta$; if mass flow erosive burning is present set $a' = a\theta$, and n' may be taken from Eq. 10-4.

It is most convenient to fit a tangent line to the burning rate curve at the equilibrium burning rate. This expression for the burning rate is

$$\frac{r}{r_e} = (1 - n') + n' \frac{p}{p_e}$$

(10-8)

Utilizing the equilibrium condition (Eq. 10-6), Eq. 10-5 may be written

$$\frac{dp}{dt} = \frac{\Gamma^2 c^*}{K} \frac{p_\circ}{l_\circ} \left(\frac{r}{r_\circ} - \frac{p}{p_\circ} \right)$$

From Eq. 10-8 this may be written

$$\frac{dp}{dt} = \frac{\Gamma^2 c^*}{K l_\circ} (1 - n')(p - p_\circ)$$

and from a differentiation theorem and Eq. 7-6

$$\frac{d}{dt}(p - p_\circ) = -\frac{p - p_\circ}{\tau} \tag{10-9a}$$

$$\tau = \frac{K l_\circ}{\Gamma^2 c^*} \frac{1}{1 - n'} \tag{10-9b}$$

$$\frac{dl_\circ}{dt} = r, \qquad l_\circ = \frac{V_\circ}{S_b} \tag{10-9c}$$

Although $l_\circ$ varies with time, this variation is so slow that a constant value may be assumed. With this assumption, integration of Eq. 10-9a yields:

$$(p - p_\circ) = (p - p_\circ)_0 e^{-t/\tau} \tag{10-10}$$

where the subscript $_0$ denotes initial value.

The pressure, if disturbed, approaches the equilibrium value exponentially. Values of τ range from 0.01 sec to 0.50 sec. At startup, Eq. 10-10 may be applied and yields a pressure-time curve resembling that of Fig. H,10. Recovery from a high pressure disturbance is indicated by the dotted lines, and the effect of variation of $l_\circ$ is shown on the "exact" curve.

Burnup. At the end of burning, the combustion chamber remains filled with high pressure gas. If the variation of gas pressure throughout the volume of the combustion chamber is neglected, then the pressure time variation, as the residual gas is exhausted, may be easily computed.

Assuming the adiabatic relation for the residual gas, and recalling that c^* is proportional to the square root of the temperature, the residual mass m in the chamber volume v obeys:

$$\frac{dm}{dt} = v_\circ \frac{d\rho}{dt} = -A_t \frac{p}{c^*} \tag{10-11a}$$

$$c^* = c^{*0} \left(\frac{\rho}{\rho^0} \right)^{\frac{\gamma - 1}{2}} \tag{10-11b}$$

$$p = p^0 \left(\frac{\rho}{\rho^0} \right)^{\gamma} \tag{10-11c}$$

The superscript ^{0} denotes initial values. Combination of these equations yields

$$\frac{dx}{dt} = -\frac{1}{\tau}(x)^{\frac{\gamma+1}{2}} \tag{10-12a}$$

$$\frac{1}{\tau} = -\left(\frac{1}{m}\frac{dm}{dt}\right)^0 = \frac{A_t p^0}{\rho^0 v_c c^{*0}} \tag{10-12b}$$

$$x = \frac{\rho}{\rho^0} \tag{10-12c}$$

Integration yields

$$x = \frac{1}{\left[1 + \left(\frac{\gamma-1}{2}\right)\frac{t}{\tau}\right]^{\frac{2}{\gamma-1}}} \tag{10-13}$$

and the adiabatic relation

$$\frac{p}{p^0} = \frac{1}{\left(1 + \frac{\gamma-1}{2}\frac{t}{\tau}\right)^{\frac{2\gamma}{\gamma-1}}} \tag{10-14}$$

Neglecting the degradation of thrust coefficient with decreasing pressure, the impulse is given by

$$I = \int_0^\infty C_F p A_t dt = C_F p^0 A_t \int_0^\infty \frac{dt}{\left(1 + \frac{\gamma-1}{2}\frac{t}{\tau}\right)^{\frac{2\gamma}{\gamma-1}}}$$

$$I = \frac{2}{\gamma+1} F^0 \tau \tag{10-15a}$$

$$F^0 = C_F p^0 A_t \tag{10-15b}$$

None of these expressions are valid after the pressure falls to the exhaust nozzle critical pressure, but the impulse estimate should not be very sensitive to this approximation.

Merits of thin-webbed grains. Very thin-webbed grains may be employed to reduce the effect of temperature sensitivity of a propellant [17]. Consider Eq. 10-10 applied to startup conditions, but with $t \ll \tau$. This gives

$$p = p_e \frac{t}{\tau} \tag{10-16}$$

It is assumed that the web of the grain is so thin that it is burned up before the equilibrium pressure is reached so that Eq. 10-16 is applicable to the entire burning period.

The burning stops at the time t_b, the web burns through, and at that time the peak combustion chamber pressure p_b is reached. Erosive effects are usually a minimum for thin-webbed motor arrangements, so that one may take Eq. 5-3 for the burning rate. Employing Eq. 10-16 for the pressure in the burning rate expression, the burning time t_b is determined from

$$\frac{w}{2} = \int_0^{t_b} r\,dt = \int_0^{t_b} a\left(\frac{p}{1000}\right)^n dt = \frac{r_e\tau}{n+1}\frac{t_b^{n+1}}{\tau} \tag{10-17}$$

where w is the web thickness, and r_e the burning rate corresponding to the equilibrium pressure. From Eq. 10-16 and 10-17 the peak pressure p_b at t_b is found:

$$p_b = p_e\left(\frac{n+1}{2}\frac{w}{r_e\tau}\right)^{\frac{1}{n+1}}$$

or

$$p_b = \left(\frac{n+1}{2}\frac{w}{r_e\tau}p_e\right)^{\frac{1}{n+1}}(p_e)^{\frac{n}{n+1}} \tag{10-18}$$

From the area ratio curve (Eq. 10-6) and the value of τ (Eq. 10-9), this may be written

$$p_b = \left(\frac{1-n^2}{2}\Gamma^2\frac{2}{l_e}\rho_pc^{*2}\right)^{\frac{1}{n+1}}(p_e)^{\frac{n}{n+1}} \tag{10-19}$$

None of the terms in parentheses depend upon propellant temperature. Therefore

$$\frac{1}{p_b}\left(\frac{dp_b}{dT_p}\right)_K = \frac{n}{n+1}\frac{1}{p_e}\left(\frac{dp_e}{dT_p}\right)_K \tag{10-20}$$

For a propellant with n about 0.6, the thin web peak pressure will vary only 0.37 as much with temperature as the equilibrium pressure. Thin web arrangements are sometimes employed to reduce the temperature sensitivity of rockets.

Resonant burning. There is one peculiar phenomenon encountered with unrestricted burning grains, particularly on the interior of cylindrical perforations, which is sometimes referred to as resonant burning. The burning process becomes unstable. Charges, which have had the burning stopped part way through the combustion process, exhibit a wavy irregular burning surface (very crudely characterized by a wavelength and amplitude of $\frac{1}{8}$ inch in some cases), and sometimes the charge is split apart. The effect on rocket operation is to produce sudden high pressure peaks which may burst the motor, loss of propellant which affects range, and general erratic behavior which reduces accuracy.

The theoretical development of this problem is not attempted here. Various studies ([10; 11; 12, pp. 893–906; 13, pp. 29–40; 14; 15; 16] and II,M) have clarified the fundamental character of the phenomenon. Practical methods to overcome this difficulty [10] consist in placing a rod of nonburning material through the perforation (a "resonance rod," steel will do), making the perforation irregular in shape rather than circular, and by drilling radial holes through the web of the grain at frequent intervals. The diameter of the radial hole need not be larger than 0.4 times the diameter of the central hole, one hole is as good as several at any longitudinal section, the axes of the holes may be coplaner but are usually spiraled around the grain to promote smoother gas flow, and the longitudinal separation between holes may be uniform but the magnitude of the separation must usually be found by experiment. Cooler and slower burning propellants exhibit resonant burning effects to a smaller degree, and for many such propellants this difficulty does not arise. Solid particles in the exhaust gases act to suppress or eliminate resonance.

Chuffing. Sometimes when a rocket motor is fired with a very large exhaust nozzle throat opening, it burns in a series of sudden pressure rises (chuffs) separated by intervals of low pressure combustion. The time interval between chuffs may amount to a few seconds to half an hour depending upon the arrangement. This behavior is usually attributed to essential stopping of combustion by an ejection of the hot combustion gas during a chuff, followed by gradual reignition from hot fittings in the motor or by hot residue from the propellant (if it is not a clean burning material). The time interval between chuffs is sometimes very regular, and the explanation may be more complex.

In the case of artillery rockets, chuffing may be dangerous. The first pressure rise may serve merely to push the rocket off the launcher, and subsequent pressure pulses may then propel the rocket around in an unpredictable manner. Even with normal rocket arrangements, chuffing may occur at low propellant temperatures.

Gas generation. Solid propellants may be employed to operate pneumatic machines. Examples are: aircraft engine starters, pilot ejectors, and fluid ejection (fire-fighting equipment). The stability and temperature sensitivity requirements depend upon the flow characteristics of the system. For example, if a solid propellant is employed to force a liquid from a tank through an orifice, the flow rate will be proportional to the square root of the pressure, rather than directly proportional to the pressure as in a rocket motor. The stability requirement is then that the burning rate index n be $n < \frac{1}{2}$, and the amplification factor for temperature sensitivity $(\frac{1}{2} - n)^{-1}$.

Actually in arrangements of this type, propellant gases are emitted into a region with cold walls, very large in area compared to the burning surface area. The cooling and condensation of gas at these surfaces results in a pressure reduction which is roughly proportional to the pressure. The net effect is to make such an arrangement about as stable as a rocket.

CHAPTER 3. SOLID PROPELLANTS

H,11. Composition and Preparation. Solid propellants may be classified into three categories: homogeneous, heterogeneous, and molecular. General comments on these types and reference to selected compositions are made in this article.

Homogeneous Propellants.

Composition of double-base powders. The most common homogeneous propellant is popularly called the "double-base powder" which is a colloid formed by gelatinizing nitrocellulose with nitroglycerin, both of which are self-combustible. Nobel patented this process to make smokeless powder in 1888. Many modifications of double-base powders have been made which have varied to a large degree both their mechanical and ballistic properties. A high nitrogen content in the nitrocellulose and a high percentage of nitroglycerin each tend to increase the energy and burning rate of the propellant. A high nitrocellulose content provides high strength propellants. The source of nitrocellulose is important; for example, when

Table H,11a. Representative composition of extruded and cast double-base propellants.

Ingredients	Extruded JPN ballistite	Experimental casting propellant
Nitrocellulose (13.25 per cent N)	51.5	
Nitroglycerin	43.0	37.7
Diethyl phthalate	3.25	
Ethyl centralite	1.0	1.0
Potassium sulfate	1.25	
Carbon black*	0.2	0.3
Candelilla wax*	0.08	
Dimethyl phthalate		14.0
Nitrocellulose (12.6 per cent N)		47.0

* Added for extruded propellant.

it is made from cotton linter it provides better physical properties than when it is based on wood pulp. In addition to the nitrocellulose and explosive plasticizer, these propellants contain a stabilizer such as diphenylamine or centralite, auxiliary plasticizers such as phthalate esters, triacetin, and dinitrotoluene, and other compounds of this nature which contribute to lower the flame temperature of combustion. For suppressing flash and to promote smooth burning at low temperatures, one or two per cent of a potassium salt has been helpful. Opacifiers such as carbon black and nigrosine dye have been found necessary for the high energy powders to prevent radiant energy from being transmitted into the propellant, where a minute flaw would absorb sufficient radiation to cause the propellant to ignite internally and to break up under the resulting gas pressure.

The composition of two double-base propellants which have been prepared by different methods—as described in the subsequent paragraphs—is given in Table H,11a.

Methods of preparation. There are three different processes by which double-base propellants are made. These are *solvent extrusion, solventless extrusion,* and *casting.*

In the solvent extrusion process, a volatile solvent is added to the mixture of nitrocellulose and nitroglycerin and the whole mix is stirred. The solvent swells the nitrocellulose and, with the breakdown of the fibrous structure by the mechanical agitation, a soft colloidal paste results, which is extruded through dies to form the desired shape and size and then cut to the desired length. The resulting charges are dried to remove the solvent. The advantages of the solvent process are that the solvent desensitizes the explosive mixtures to reduce manufacturing hazards and the charges are physically stronger than those made by the other processes. The disadvantages are that the propellant cracks during drying if the maximum propellant thickness exceeds about ½ inch and that control of shape and dimensions is difficult, because of the shrinkage which accompanies solvent removal.

In the solventless process the nitrocellulose-nitroglycerin colloid is formed by mixing these ingredients first in a water slurry and then transferring them to a heated roller mill. The propellant is removed in sheet form, either rolled like a carpet or flaked into small pieces and inserted into a press, the barrel of which is heated between 115 and 130°F, and then extruded at pressures ranging from 4000 to 6000 lb/in.² The temperature of the extruded propellant is usually about 30°F less than the press barrel temperature. The space between the ram and the propellant is initially evacuated to prevent formation of bubbles and "pinholes" in the extruded powder. The extruded charge is finally annealed to relieve stresses. The advantages of the solventless process is that large charges can be made (the limitation being the size of the press) and that exact control of size and shape is possible. The disadvantages are physical properties inferior to solvent-processed propellant, heavy and costly equipment, and hazardous operations. However, most of the rocket propellant used in World War II was made by the solventless process.

HETEROGENEOUS PROPELLANTS.

Composition. These propellants generally consist of an organic fuel binder and crystalline oxidizers which have been intimately mixed together. A large number of propellants of this type are possible because of the many fuels and oxidizers which are available. Thermoplastic mixes such as asphalt-oil and ethyl cellulose-castor oil, and thermosetting materials including several synthetic resins and elastomers have been feasible as fuel binders. Perchlorates and nitrates have been used most extensively as oxidizers, among the most common being ammonium nitrate, ammonium perchlorate, potassium perchlorate, and potassium and sodium nitrates. For example, a product of reaction of potassium perchlorate is potassium chloride, a white solid; hydrogen chloride from ammonium perchlorate condenses in a moist atmosphere to produce a white fog. Nevertheless, because of the simplicity of processing and the wide range of burning rates, these propellants can compete with smokeless double-base powders in many applications. The composition of two propellants which have been made by the casting and molding processes is given in Table H,11b.

Table H,11b. Composition of two heterogeneous propellants.

Ingredients	Castable asphalt-perchlorate	Molded composite
Asphalt	16.8	
Oil	7.2	
Potassium perchlorate	76.0	
Potassium nitrate		36.4
Ammonium picrate		54.6
Calcium stearate		3.6
Ethyl cellulose		5.4

Preparation. Heterogeneous propellants are prepared by *casting* and *molding* techniques.

In the cast process the propellant is usually fluid at some stage of the preparation so that it can be poured into molds of any desired size and conform to any shape. The fuel is usually the ingredient that provides fluidity. Thermoplastic fuels are usually originally solid, are melted to be mixed with the oxidizer, and are then cooled to a solid after casting [46]. Propellants using thermoplastic fuels are cooled gradually to avoid severe thermal stresses which cause cracks or pull-away from molds. Thermosetting fuels are usually liquid at normal temperatures and are solidified by chemical polymerization. The polymerization of thermosetting propellants is usually exothermic, so that the reaction has to be controlled to avoid high thermal stresses from this source and to prevent autoignition of propellant; therefore, control of the amount, chemical purity, particle size (if solid), and dispersion of the catalysts, which promote polymerization and control of the temperature at which the polymerization occurs, are essential.

Oxidizers are pulverized to very fine powders before being incorporated into the fuel. The distribution of oxidizer particle size often ranges from 5 to 40 microns. The classification of oxidizers on a large scale into narrow particle size ranges is an operation which has not been satisfactorily controlled and is a good subject for research. Both the mixing and burning properties of solid propellants depend largely upon the particle size distribution of the oxidizer. Many of the oxidizers are hygroscopic and the control of the humidity during processing is mandatory.

Considerable work has been done in an attempt to relate the properties of solid propellants with viscosity of the fuel, the volume concentration of the oxidizer, and the particle size distribution of the suspended oxidizer. The first work relating the viscosity of suspension with concentration of the suspended materials is due to Einstein. However, the Einstein equation was only applicable to extremely dilute suspensions. Later Vand [47] developed equations for the viscosity of suspensions as a *function of concentration which held to about 30-volume-per cent solids.* Since the Vand equations do not take into account non-Newtonian flow, they are not applicable to the high concentration commonly encountered in composite solid propellants. However, they are useful for first order approximation. Although considerable effort has been expended toward attempting to derive mathematical formulas to describe the behavior of suspensions as concentrated as those encountered with solid propellant, no suitable equations have been developed to date.

In addition to the concentration of the suspended particles, the particle size distribution of the particles is extremely important. Dalla Valle [48, pp. 100–122] has written a comprehensive book on the effect of varying particle size distribution on various properties (such as bulk density) which influence the rheology of suspensions. Furnas [49] and Mooney [50] have worked on the viscosity of suspension of spheres. However, since the oxidizers are not entirely spherical in shape, none of the equations for particle size distribution as a function of viscosity hold rigidly. The results obtained from spherical particles, however, can be used as a first approximation of obtaining optimum results with solid propellants.

Settling of solid propellants prior to solidification presents a problem since any significant sedimentation results in nonhomogeneous castings. The basic equation for the sedimentation velocity of particles in suspension is given by Stokes' law. Stokes' law does not, however, hold for concentrated suspensions, and alternate equations have been developed by Steinour [51] and Hawksley [52].

Compromises usually must be made between viscosity of the liquid fuel and particle size distribution of the oxidizer, in order to obtain the proper mixture ratio for ballistic performance and yet retain adequate mixability and castability without settling. In practice, a mixture ratio of 80 per cent by weight of oxidizer and 20 per cent by weight of fuel is about the maximum feasible for a system having: (1) an oxidizer having an approximate specific gravity of 2.0, (2) the oxidizer being blended in the ratio 70:30 per cent by weight with each fraction having a particle size of 50:5 microns, respectively, (3) a fuel with an approximate specific gravity of 1.0, and (4) the fuel having a viscosity of about 1000 centipoises.

For the mixing operation it is generally necessary to heat or cool the propellant mixture and often necessary to evacuate the air from it to provide a nonporous propellant. Mixers with paddle-type agitators and double counter-rotating differential speed blades have been used most extensively for these cast propellants.

The chief advantages of heterogeneous cast propellants are that the equipment cost is very low and the size of propellant charge appears to be unlimited. As a matter of interest, some of these propellants are the only ones existing which do not depend upon nitric acid. The big disadvantage is that most of these propellants produce smoke.

In the pressure-molding process of heterogeneous propellants, pressures of the order of 10,000 lb/in.² are used to consolidate the propellant. Consequently, heavy metal parts and large presses are often required which are cumbersome and make this process much more expensive than the cast process. Also, the charge size may be limited by the size of the press. However, sometimes molding is desirable for special propellants which could not be made by the cast process, an example being a material with nearly all of the ingredients crystalline and only a small amount of fuel binder permissible.

Two mixing methods have been popular for molded heterogeneous propellants. One method makes use of the intensive mixing of an edge-runner mill; the other, a differential roller mill. A propellant system consisting of ammonium picrate, sodium nitrate, and a small portion of resinous binder—popularly called composite propellants—has been successfully processed using the edge-runner mill. The resultant propellant mixture is a powdery product that is formed by compression, molding at about 4000 to 8000 lb/in.². The molded charges are cured for about 4 days at 175°F. A charge length about equal to the diameter has been the most satisfactory for composite propellants and charges with an l/d greater than 1 are made by cementing two or more charges together.

When using a differential roller mill the oxidizers and miscellaneous additives are blended into the fuel and the final mixture of propellant is removed from the roll as a sheet, similar to the solventless double-base powder process. If the fuel is thermoplastic, it may be molded under fairly low pressure directly into and bonded to rocket motors. These charges are of puttylike consistency and are often subject to plastic flow, but when the propellant is restrained and forced by thermal changes to expand or contract, dimensional changes in the direction of unrestraint usually do not set up stresses large enough to cause cracking. If the fuel is thermosetting, it can be forced into a mold which is subsequently locked and put into an oven where the temperature initiates polymerization. As the temperature expands the confined propellant, the propellant fuses to form one solid piece. After polymerization, the propellant is cooled and removed from the mold.

Molecular Propellants. Molecular propellant is the name given to a propellant consisting of one basic molecule. The single-base smokeless gun powder cellulose nitrate is an example of such a propellant, but it does not burn properly at pressures below 5000 lb/in.^{2}, the regime in which most rocket motors are designed. The molecular propellant would be ideal for processing if it were liquid at normal temperatures and could be polymerized, when desired, to a solid. Any size and shape of charge could be made with a minimum of effort. No published information is available on the status of the research with molecular propellants, but it is apparent that much work remains to be done before this type of propellant becomes practical.

Restrictors. After the propellant is cast to size and shape there are usually many cutting, drilling, and sawing operations before the charge is a finished product ready to be inserted into a rocket motor. One of the most important operations performed is that of applying a restrictor, which is a noncombustible layer applied to a portion of the propellant surface where burning is not desired. The process of applying a restrictor is called inhibiting. Many blowups of rocket motors have been due to failure of the restrictor.

Cellulose acetate has been satisfactory for restricting double-base propellants. For extruded double-base charges, the cellulose acetate—usually molded to a given size or cut to the desired shape from sheet stock—is often cemented to the propellant. For the cast double-base powders, the cellulose acetate in tube form often serves as the casting mold as well as for the restrictor.

The fuel with an inert filler substituted for the oxidizer has been found to be a very satisfactory restrictor for many of the heterogeneous propellants. For cast heterogeneous propellants the fluid restrictor may be applied in a thin layer ($\frac{1}{32}$ to $\frac{1}{16}$ inch) directly to the rocket motor wall by draining or spinning methods; the liner bonds to the motor wall when it is solidified by cooling or polymerization. The propellant then adheres to the liner after it is poured into the motor and solidified. In a few cases, the propellant itself may bond satisfactorily to the metal of the motor wall, thus eliminating the need for a restrictor. Heterogeneous propellants are often cast separately from the motor, however, and in this case the mold may serve as the restrictor, or the charge may be inhibited by dipping it into a pool of restriction material; often, additional wrappings of tape over the restrictor are found desirable. For molded heterogeneous propellants, restrictors may be made of special materials which are bonded in shaped pieces to the charges, for example cellulose acetate is bonded to double-base powders. For some elastomeric propellants, placing sheets of fuel at desired places on the propellant charge prior to molding has proved to be simple and satisfactory.

H,12. Propellant Properties. The efficiency and usefulness of a solid propellant rocket depend to a large extent upon the properties of the propellant. The combustion, mechanical, and physical properties of the propellant must be known in order to design and properly use such rockets. A discussion of these properties with reference to specific propellant systems is presented here.

COMBUSTION.

Basic propellant characteristics. By examination of the burning rate and equilibrium pressure relationships given by Eq. 5-1, 5-2a, 5-2b, 5-3, 7-13 and 9-1, it is evident that the basic characteristics which define the combustion of a given propellant are a, n, k or k_V and V_0, c^*, and ρ_p.

The burning rate coefficient a is the propellant characteristic which varies most significantly with propellant temperatures, so that it is an extremely important factor in rocket motor design. The sensitivity of burning rate to propellant temperature, T_p, change may be expressed as:

$$\pi_r = \frac{1}{r}\left(\frac{\partial r}{\partial T_p}\right)_p = \frac{1}{a}\frac{\partial a}{\partial T_p} = \frac{1}{T_1 - T_p}$$

where the last result follows from Eq. 5-2. Recall that T_1 is an experimental property of the propellant. π_r is defined as the *temperature sensitivity of burning rate at constant pressure.* The values of T_1 for some different propellant systems are approximately as follows: 400 to 600°F for double-base propellants; 500 to 600°F for propellants utilizing ammonium nitrate or ammonium perchlorate for oxidizers; greater than 700° F for propellants utilizing potassium perchlorate for oxidizer. It is evident that the constant T_1 should be high compared to the useful range of the propellant temperatures, T_p.

In addition to affecting the burning rate as the propellant temperature changes, the burning rate coefficient partly determines the magnitude of the burning rate. Since a choice of burning rate is convenient in rocket motor design, it is desirable to be able to vary the constant a over a wide range.

The burning rate exponent n is equally as important as the burning rate coefficient because of its influence on the variables which determine the equilibrium chamber pressure of a rocket motor. From Eq. 7-13 these variables are raised to be $1/(1-n)$th power. As $n \to 1$, the pressure is extremely sensitive to small design changes. For example, if $n = 0.95$ and the nozzle throat area decreased by 1 per cent, the pressure would increase 73 per cent. *The temperature sensitivity of pressure at constant K value* is very important in rocket design. From Eq. 10-3,

$$\pi_p = \frac{1}{p}\left(\frac{\partial p}{\partial T_p}\right)_K = \frac{1}{1 - n'}\left(\frac{\partial r}{\partial T_p}\right)_p = \frac{1}{1 - n'}\pi_r$$

where n' is defined in Eq. 10-4. It is evident that for ideal operating characteristics the burning rate exponent should approach zero.

Constant burning rates ($n = 0$) have been obtained over limited pressure regions for both heterogeneous and homogeneous propellant regions. Small amounts of particular additives produce the constant burning rate effect for both the double-base and heterogeneous propellants. When the propellant gases flow parallel to the burning surface of a solid propellant, the burning rate of the propellant at a given pressure is increased (see Eq. 5-1 and 5-2). The proportionality factor k is called the erosive burning constant. Due to erosive burning the additional mass flow increases the rocket motor chamber pressure according to Eq. 9-1. The effect of erosive burning on the pressure-time characteristics of a rocket motor is indicated by Fig. H,12a.

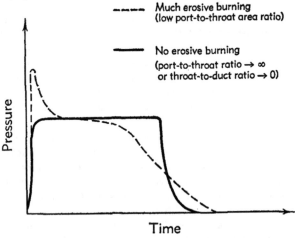

Fig. H,12a.   Effect of erosive burning on the pressure-time curve of a rocket motor. Charge design conditions identical except for throat-to-duct area ratios.

The erosive burning constant can be determined experimentally by stopping the burning of a radial-burning solid propellant charge and measuring the distance burned versus the length of charge—which has been found to be nearly a linear relationship—then calculating from geometry the average mass flow density at any desired point along the charge for the time of burning, and finally making the calculation for the erosive constant using the basic relation of Eq. 5-1 and 5-2. The propellant at the front end of the charge is assumed to burn under nonerosive conditions and any significant pressure drop along the charge has to be taken into account.

In a given propellant system using different ratios of ingredients, the erosive constant may vary by a factor of 5. In general, it has been found that low energy propellants have high erosive constants, and high energy propellants have low erosive constants. The resistance of the propellant to the shear forces exerted on it by the gases may also be an important factor. Values of k usually range from nearly zero to 0.7 in.^{2} sec/lb.

The characteristic velocity c^* is a measure of the intrinsic energy of the propellant, and is determined by the thermodynamic properties. For an idealized propellant model, these are T_0, γ, and $\mathfrak{M}$ (see Sec. G). It is essentially independent of propellant temperature and varies little in the pressure range normally encountered with rockets. This insensitivity of energy to initial propellant temperature is in contrast to the possible large dependency of the rate processes upon temperature. It follows that the total impulse of a rocket unit is nearly always the same, but that the rate of expenditure of energy varies with propellant temperature.

It is apparent that a high c^* is desirable. However, since $c^* \sim \sqrt{T_0/\mathfrak{M}}$, a high value of T_0 may cause excessive heating of parts of the rocket motor that are exposed directly to the hot gases. Nozzle erosion, in particular, is severe with conventional materials when T_0 approaches or exceeds 4000°F and the duration of rocket motor operation is more than a few seconds. For this reason, it may be more desirable to increase c^* by reducing $\mathfrak{M}$ than to increase T_0. In general, c^* values greater than 4000 ft/sec are easily attained for solid propellants, but values significantly exceeding 5000 ft/sec are not likely to be common in the foreseeable future.

The product of c^* and ρ_p should be high to provide a high impulse per unit volume. A high energy propellant of low density may not provide an efficient rocket motor because of the additional weight of the rocket motor chamber necessary to enclose the volume occupied by the propellant. The density should be high. The density of most solid propellants ranges between 0.05 lb/in.³ and 0.07 lb/in.³

The products of combustion are important because they determine smoke, toxicity, and corrosiveness of the exhaust. For example, potassium chloride from potassium perchlorate gives a dense white exhaust, carbon monoxide is very toxic, and hydrogen chloride from ammonium perchlorate is very corrosive to materials such as iron. Carbon monoxide reacts vigorously with steel nozzles to form iron carbonyl when small amounts of sulfur are present in the exhaust.

Special combustion characteristics. There are some special combustion properties which sometimes impose major limitations on rocket design. Such characteristics as the combustion limit, pressure limit, temperature limits, and nonuniform burning properties are discussed in this article.

If a number of rockets, identical except for the exhaust nozzle throat diameter, are fired, a graph such as Fig. H,12b can be plotted showing the chamber pressure as a function of the throat diameter [42]. It is found, as might be expected, that the pressure decreases as the throat diameter increases. However, when a certain throat diameter is exceeded, the chamber pressure is found to be far below the pressure predicted from an extrapolation of the high pressure portion of the curve. The pressure corresponding to this critical thrust diameter is called the *combustion limit*.

If a number of units are fired with this same large throat diameter the chamber pressure varies erratically from unit to unit. Finally, if the exhaust nozzle throat is made large enough, individual units do not burn continuously, but in an irregular manner with a chugging noise. This last phenomenon is called "chuffing."

Thus it appears that a given propellant cannot be used at an arbitrarily low pressure, and the designer of a rocket unit must assume chamber pressures above the combustion limit if reproducible performance from unit to unit is to be obtained. When low over-all rocket weight is

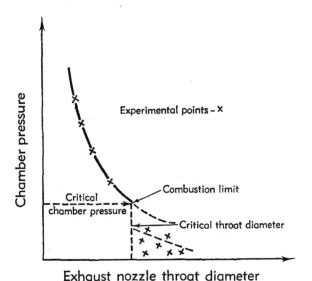

Fig. H,12b. Illustrating the combustion limit.

desirable, a high combustion limit is a serious disadvantage, since the weight of the walls of the combustion chamber is essentially proportional to its volume and to the design chamber pressure.

The combustion limit of some extruded double-base propellants is about 500 lb/in.²; that of cast asphalt-potassium perchlorate propellants about 1000 lb/in.² The composite propellants have a combustion limit not exceeding 100/in.² The combustion limit of the straight nitrocellulose propellants frequently used in guns exceeds 5000 lb/in.²; therefore such propellants are not suitable for rockets.

Some propellants may safely be used only below a critical chamber pressure, this pressure being called the *pressure limit*. If the pressure limit is exceeded, the propellant burns in a violent and unpredictable manner. Brittle propellants with a granular structure are particularly subject to this effect. For most of the commonly used rocket propellants the pressure limit exceeds 5000 lb/in.², and therefore it is not a problem in rocket unit design.

Special limitations on the temperature range within which a rocket may be used are sometimes introduced by a change of mechanical properties of the propellant with temperature. Certain types of propellants, those having thermoplastic fuels, may soften and deform sufficiently when stored at high temperature so that an abnormally large burning area is exposed. Upon ignition, the resulting high chamber pressure may cause failure of the unit. Some double-base propellants may soften sufficiently at high temperatures so that the large pressure gradient along the charge, which is present in radial-burning units, may cause the charge to break up, exposing a large burning area and leading to failure of the unit. At

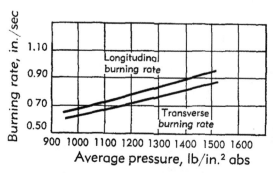

Fig. H,12c. Longitudinal and transverse burning rates of a double-base extruded charge.

sufficiently low temperatures, some propellants may become so brittle that they shatter upon ignition of the charge. The large burning area exposed in a failure of this kind may lead to a violent explosion. Improvement of *temperature limits* is a continuing task in propellant development.

Propellant charges may not always have uniform burning characteristics. For example, the burning rate in the longitudinal direction of extruded double-base charges differs from the rate in the transverse direction, an example of the difference being shown in Fig. H,12c [53]. As a consequence, these propellants, burning in a radial direction, burn differently than those burning from one end. Also, the burning rates depend upon whether the propellant is extruded in a direction parallel or perpendicular to the propellant sheet obtained from the mixing rolls. Presumably the orientation of the colloidal structure affects the burning rate.

METHODS FOR DETERMINING COMBUSTION CHARACTERISTICS. After it has been determined by calculation that a given propellant system has acceptable energy (c^*), the burning properties have to be evaluated. The burning rates of laboratory quantities of propellant are often determined by preparing pencil-size specimens and then measuring the time required for given lengths to burn from one end to the other at different pressures and temperatures; this method is referred to as the *strand burner method* [*31*]. The strand burner has proved to be very useful but has limitations, because the pressurizing gas (usually nitrogen) never duplicates the combustion gases upon which the burning characteristics are dependent. Sometimes, propellants which will not burn within a given pressure range in the strand burner will burn satisfactorily at the same pressure in rocket motors.

A simple method for determining the rocket propellant characteristics over a wide pressure range is to burn an internal-burning tubular charge which is restricted on the ends in a rocket motor using high port-to-throat ratio[6] (negligible erosive burning). The weight of propellant burned at any time is proportional to $\int_0^t p \, dt$; therefore the distance burned at any time can be calculated from the weight burned and geometry. As a consequence, the burning area corresponding to the pressure at any time can be calculated; then the burning rate exponent can be determined by the relation $p \sim A_b^{1/(1-n)}$. Since the calculation of ρ_p and c^* is straightforward, the burning rate coefficient can be determined from the relation

$$p = \left(\frac{c^* a \rho_p K}{g}\right)^{\frac{1}{1-n}}$$

This method of determining the burning characteristics is very useful for finding the exact limits of "constant burning rate" regions, because the change in burning rate exponent is shown immediately by the pressure-time curve (see Fig. H,12d). The method depends upon a constant throat area during burning, which is difficult to obtain in practice when the throat area is small (say less than 0.2 in.); therefore it is applicable to charges which are usually larger than those made in the chemical laboratory.

STORAGE STABILITY. The mechanical and internal ballistic properties of a propellant may be dependent upon storage time, storage temperature, and humidity.

The storage time permissible for double-base propellants is strongly dependent upon temperature because these propellants decompose chemically. The chemical decomposition is autocatalytic, and stabilizers such as diphenylamine have to be added to neutralize the catalytic effect of the decomposition products; otherwise these propellants may autoignite. This decomposition leads to the formation of gas which tends to fissure the charges, particularly those having thick sections of propellant. Another problem has been the exudation of the nitroglycerin into the

---

[6] The term port is synonomous with duct in rocket terminology. The port-to-throat ratio, $A_d/A_t = H$, is the reciprocal of the throat-to-duct (or throat-to-port) ratio: $A_t/A_d = J = 1/H$. In Art. 8 on theory, J was used since it ranges from 0 to 1.

restrictor to such a degree that the restrictor burns readily. The stability of the double-base propellants is being constantly improved, but storage at temperatures exceeding 140°F is inadvisable.

The composite propellants do not decompose chemically on prolonged storage but, in an atmosphere of high humidity, the sodium nitrate absorbs moisture and the charges containing it become soft and weak. Propellants containing ammonium nitrate are very hygroscopic, and the phase changes of this material make its use impractical for propellants

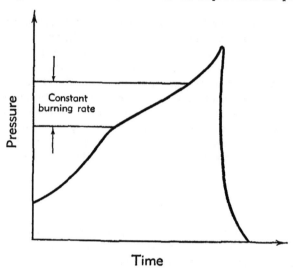

Fig. H,12d. Example of the constant burning rate pressure region of the burning rate for a propellant, as determined from a rocket motor test employing an internal-burning tubular charge.

stored and used over a wide temperature range unless the fuel binder can accommodate the volume changes.

Propellants containing volatile ingredients lose weight on prolonged storage at high temperatures, resulting in changes for both mechanical and burning properties.

The heterogeneous propellants with nonhygroscopic stable oxidizers and nonvolatile fuels usually have excellent storage stability. Satisfactory storage for longer than a year at temperatures exceeding 160°F is feasible.

MECHANICAL PROPERTIES. Constancy of mechanical properties over a wide temperature range is desirable so that brittleness at low temperature and plastic flow at high temperatures do not limit the use of the rocket propellant to a narrow temperature range.

In withstanding flight accelerations, the mechanical properties required of a rocket propellant depend upon the particular charge design being used. For illustration, three charges: end-burning, internal-burning, and tubular-burning (burning on both outside and inside) are considered, as shown in Fig. H,12e. If rocket motors containing these charges are accelerated at y times gravity, the following situations exist:

For the end-burning charges. The opposing forces are the acceleration "pressure" $y\rho_p l_p$ acting to throw the charge rearward and the chamber pressure p holding the charge forward. If the thrust which accelerates the motor is from the end-burning charge only, the pressure force is

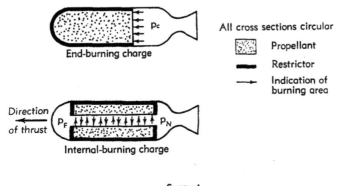

Fig. H,12e. Various propellant arrangements discussed in connection with the effect of forces on the charge.

always greater than the acceleration because in the expression

$$\frac{p}{yl_p\rho_p} = \frac{p}{\dfrac{F}{W_T}\dfrac{W_p}{S_b}} = \frac{pS_bW_T}{C_Fp A_t W_p} = \frac{S_bW_T}{C_FA_tW_p}$$

the numerator is greater than the denominator, since

$$\frac{1}{C_F} > \frac{1}{2}, \quad \frac{S_b}{A_t} > 100, \quad \frac{W_T}{W_p} > 1$$

In the above formulas S_b is the burning surface of the propellant, W_T the total weight of the rocket motor, W_p the weight of the propellant, and C_F the nozzle thrust coefficient. The main requirement for this end-burning charge is that the propellant be "plastic" enough or have enough elongation to change dimensions without cracking as required by the rocket motor expansion under pressure. High elongation of propellant is desirable to "accommodate" the differential thermal expansion properties between the motor chamber and the propellant.

For the internal-burning charge. The forces tending to throw this charge rearward are due to the pressure drop along the charge and acceleration. These are:

Pressure drop force (see Art. 8 and Fig. H,9j)

Acceleration force $= yW_p$

Opposing these forces is the adhesion of the propellant to the motor wall or the cohesion to itself, whichever is weaker. It is evident that the adhesion strength in shear and the shear strength of the propellant are very important. For identical conditions, the stress increases as the charge becomes larger in diameter. Since this propellant is bonded to the combustion chamber, it should have good elongation to follow, without cracking, the expansion of the rocket motor under pressure and to withstand the dimensional change with temperature change. It is noted that J becomes less as the chamber wall expands, thus tending to decrease the pressure drop force.

For the tubular charge. The forces tending to throw the charge rearward are the same as for the internal-burning charge. Because the charge is supported by the cross-sectional area at the nozzle end, it must take the load in compression. It follows that the compressive strength should be high. However, the compressive strength for such a charge is not always the mechanical property, which is of prime significance [54]. Contrary to the internal-burning charge bonded to the motor wall, the gas passage becomes smaller as chamber pressure becomes higher, thus making J larger (see Art. 6). This leads to increased pressure, increased pressure drop force, and increased acceleration force. It follows that the modulus of elasticity should be high to prevent "bulging" of the charge under high compressive force. Also, at a time near the end of burning of tubular charges, a very thin propellant section is left which may buckle as a column; since the strength of a column is directly proportional to the modulus of elasticity and inversely proportional to its slenderness ratio, the significance of modulus of elasticity and design is again emphasized.

From the preceding discussion, it is evident that the most important mechanical properties may range from high elongation, high adhesive and cohesive shear strength to high modulus, and high compressive strength. It should be added that, since the mechanical properties of propellants are never constant with temperature, their variation with temperature must be considered in design studies.

Unfortunately, conventional test methods for evaluating the strength properties of materials do not give quantitative design information when applied to propellants. This is true because most propellants have complex molecular structures which do not obey conventional laws of elasticity. Properties such as tensile strength and elongation are significantly dependent upon the rate at which the test specimen is loaded and its complete past history with respect to temperature, humidity, and time. The variation of elastic modulus with propellant temperature is shown in Fig. H,6d. Regardless of the fact that the measured mechanical properties of propellants have not been usable for absolute design, knowledge of the behavior of the properties in a qualitative sense is extremely valuable for rating one propellant against another. Development of test methods from which quantitative designs can be made is a subject needing much research.

MISCELLANEOUS PROPELLANT PROPERTIES. Some additional, important properties of propellants are discussed briefly in this article.

Shrinkage. Some of the polymeric fuels become more dense as the polymer chains grow, thus resulting in shrinkage. If volatile materials evaporate from propellants, shrinkage occurs. The amount of shrinkage is useful for calculating final dimensions of charges and for estimating stresses which may result. When a propellant which may shrink is restrained from shrinkage—such as when a solid core is used for shaping the internal perforation of a charge—stresses develop in the propellant, sometimes high enough to produce cracks. Core withdrawal may be increasingly difficult as shrinkage continues.

Thermal properties. *The thermal coefficient of expansion* is required in order to allow proper tolerances between the propellant charge and the metal parts, and to estimate stresses when the propellant is bonded to the motor wall or when the propellant is free to contract around the cores used for shaping the charge.

Thermal diffusivity is needed to determine temperature distributions [55] according to the equation (circular symmetry, no axial heat flow)

$$\frac{\partial T}{\partial t} = \kappa \left(\frac{\partial^2 T}{\partial R^2} + \frac{1}{R} \frac{\partial T}{\partial R} \right)$$

where $\kappa = k/c_p\rho_p$. The thermal conductivity and specific heat are the basic thermal properties that determine the value of thermal diffusivity. Typical values of thermal coefficient of expansion, thermal diffusivity, and specific heat for propellants are 10^{-4} in./in./°F, 25×10^{-5} in.²/sec, and 0.3 cal/g°C, respectively.

The specific heat of reaction c_r. It is necessary to know this in order to find the maximum internal temperature rise possible during the polymerization of propellants.

$$\Delta T = \frac{c_r}{c_p}$$

The maximum temperature should be well below the maximum safe storage temperature for the propellant.

Ignition properties. *The ignition temperature* of a propellant varies with the time of exposure. The temperature below which ignition is unlikely to occur should be found for every propellant. Unfortunately no standard ignition method has been established, so that ignition values quoted by different laboratories should not usually be compared. In the absence of a standard method, the value of T_1 (Eq. 5-2) gives an order of magnitude of the autoignition temperature, because from this equation (which is really determined from temperature sensitivity data rather than autoignition data),

$$r \to \infty \quad \text{as} \quad T_p \to T_1$$

However, it is dangerous to assume that T_1 has any physical significance with regard to the temperature below which a propellant will ignite. Some propellants may ignite when stored at temperatures several hundred degrees below T_1. Adiabatic heating brought about by chemical decomposition may cause ignition. Great care should be utilized in interpreting ignition data: size, shape, and storage conditions of the propellant are very significant.

CHAPTER 4. DESIGN OF ROCKET MOTORS

H,13. Discussion of Requirements. The primary function of the rocket motor usually is to supply a given propulsive impulse, and the

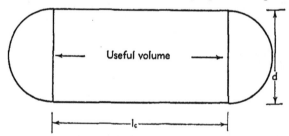

Fig. H,13a. Idealized chamber (nozzle neglected) utilized in discussing impulse-weight relationship.

designer strives to achieve the impulse requirements with a rocket motor of minimum weight. An index of efficiency often used to compare rocket motors is the impulse-weight ratio, or the total impulse divided by the total rocket motor weight. By using Fig. H,13a to depict a rocket motor, the impulse-weight ratio of a rocket motor may be shown to be [42]

$$\frac{I}{W_\tau} = \frac{I_{sp}}{\left(2 + \dfrac{d}{l}\right) \dfrac{\rho_w}{\rho_p} \dfrac{p_o}{\sigma_w} \dfrac{1}{v_e} + 1}$$

Here d is the useful volume diameter, l is the useful volume length, ρ_w is the density of the combustion chamber wall material, σ_w is the design stress for the wall material, p_o is the design chamber pressure, ρ_p is the propellant density, and v_e is the fraction of *total* volume occupied by the propellant. An impulse-weight ratio greater than 100 seconds usually represents good design. It is evident that the propellant should have a high specific impulse and a high density, that the combustion chamber should be of materials having a high strength-to-weight ratio, that the volume available for propellant be used efficiently, and that the diameter-to-length ratio be low. It appears that the pressure should be as low as possible, but I_{sp}, d/l, and v_e each vary with pressure in such a way that an optimum pressure does exist. For the majority of applications rocket motor operation is most efficient between 500 and 1500 lb/in.^{2} abs. In addition to total impulse, the time rate at which the impulse is expended is usually specified, and it is this requirement that presents the most difficult design problem because of the limitations of available burning rates. Examples of some of the conflicting requirements follow.

Tubular charges have often been used for artillery rockets for which a reasonably short burning time is desired to obtain accuracy. Also, a high velocity is necessary for penetration and range. However, these requirements are not "compatible," because a short time can be achieved only by a thin propellant web (low propellant weight) and a high pressure (heavy metal parts), both of which result in reduced velocity and range.

For a high altitude rocket, a small frontal area is desirable to minimize the aerodynamic drag; this often leads to a small diameter requirement which restricts the web thickness for radial-burning charges. A thin web limits the burning time and propellant weight in such a way as to increase the significance of aerodynamic drag, so that, again, the requirements conflict.

Since the rate of expenditure of a fixed amount of a given propellant depends primarily upon the product of burning surface and burning rate, and the burning rate is limited by basic propellant properties, the designer must use ingenuity to obtain a practical charge geometry which has the necessary burning surface. In many radial-burning designs it is difficult to maintain a constant burning surface, and, as a consequence, the pressure is not constant; in order to keep the pressure below the maximum design pressure and above the combustion limit, the allowable pressure variation (or burning surface variation) is limited.

An optimum motor design. From the preceding discussion on requirements for rocket motor design, it is evident that an optimum rocket motor exists. The optimum motor is by no means *one* motor, however, and depends entirely upon the use for which it is intended. A free flight rocket is designed differently from a rocket used for auxiliary aircraft power; or, in fact, optimum design for rockets intended for a given purpose varies according to, for example, payload. Many rocket motors have been inefficiently used in applications for which they were not primarily designed.

The question often arises, "What rocket motor will give the longest range for a given payload?" For the purpose of illustrating a solution to this problem it is assumed that the rocket is fired horizontally in a vacuum a fixed distance above the earth. The rocket having the highest burnt velocity will achieve the longest range. The burnt velocity may be expressed by

$$V_b = -gI_{sp} \ln(1 - \nu)$$

For a given propellant the specific impulse depends only slightly upon chamber pressure and therefore the propellant-to-gross weight ratio ν is the main variable.

$$\nu = \frac{W_p(1 - \sigma)}{W_g} = \frac{W_p(1 - \sigma)}{W_p + W_n + W_c + W_1 + W_m}$$

or

$$\nu = \frac{1 - \sigma}{1 + \dfrac{W_n}{W_p} + \dfrac{W_c + W_1 + W_m}{W_p}}$$

By assuming that the weight of the nozzle is proportional to the total impulse

$$\nu = \frac{1 - \sigma}{1 + \phi I_{sp}(1 - \sigma) + \dfrac{W_c + W_1 + W_m}{W_p}}$$

The velocity is greatest when ν is maximum or when $W_c + W_1 + W_m/W_p$ is minimum. Now for a given payload, pressure, chamber materials, propellant and restriction materials, duct-to-throat area ratio, and initial-to-final burning surface ratio, the value of $(W_c + W_1 + W_m)/W_p$ is solely a function of length-to-diameter ratio which can be optimized to give maximum velocity. For the purpose of illustrating the solution, a study [56] has been made by selecting two propellants of extremely different burning characteristics and by using the configuration and conditions shown in Fig. H,13b. After optimum values of l/d were obtained at various pressures, fictitious maximum ranges[7] R' were calculated by

$$R' = g[I_{sp} \ln(1 - \nu)]^2$$

and plotted vs. optimum l/d ratios and corresponding chamber pressures. The results are shown in Fig. H,13c and H,13d for the two propellants with zero payload and with a payload corresponding to one caliber of a material weighing 0.5 lb/in.[3] It is seen that the best value of l/d becomes lower as the chamber pressure and payload decrease and as the burning rate exponent n becomes larger. Of course, these curves indicate only general trends for other specified cases where different values for duct-to-throat area ratio and initial-to-final burning surface ratio are considered.

[7] The range R' corresponds approximately to the maximum range possible if the rocket had zero burning time and were fired upward at optimum quadrant angle.

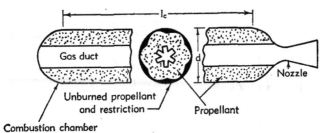

Fig. H,13b. Conditions for optimum rocket motor design study.

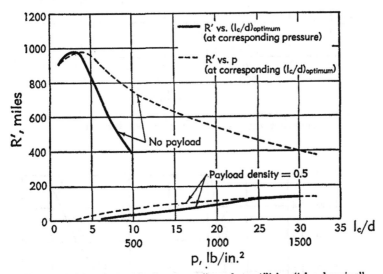

Fig. H,13c. Optimum drag-free range for rockets utilizing "slow burning" propellant with $r_{1000psi} = 0.35$ in./sec (see Fig. H,13b).

For vertical flight in atmosphere, the burning time partly controls the drag deceleration. The weight of loading per unit cross-sectional area μ also partly determines the deceleration due to drag and may be expressed by

$$\mu = \frac{\text{total weight of rocket}}{\text{cross-sectional area}} = \text{const}$$
$$\times \left[\frac{W_p}{d^3} 1 + \phi I_{sp}(1 - \sigma) + \frac{W_o + W_1 + W_m}{W_p} \right]$$

Therefore, in a study where drag and gravity are considered, the burning time, propellant weight, and diameter are important rocket motor characteristics which may entirely change the optimum l/d for maximum range. The variation of t_o/d and W_p/d^3 vs. l/d at various pressures is

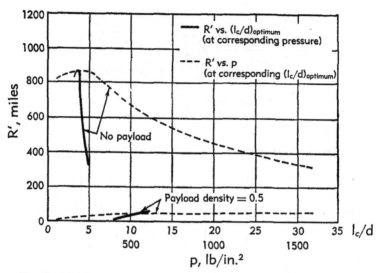

Fig. H, 13d. Optimum drag-free range for rocket utilizing "fast burning" propellant with $r_{1000psi} = 0.75$ in./sec (see Fig. H,13b).

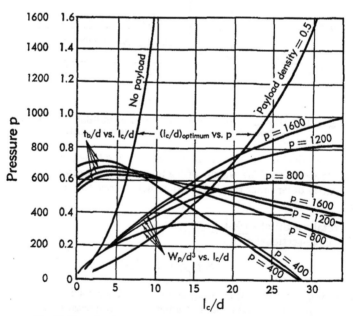

Fig. H,13e. Rocket motor characteristics for "slow burning" propellant with $r_{1000psi} = 0.35$ in./sec (see Fig. H,13b).

given in Fig. H,13e and H,13f for the study which has just been presented for horizontal flight, vacuum conditions. Within the limitations imposed, it is seen that low chamber pressure, small l/d, and large diameter each increase the burning time; the propellant weight decreases as the l/d ratio becomes less than optimum and as the chamber pressure is reduced.

With regard to cost, it may be desirable to design rocket motors with l/d's less than optimum, since it has been shown that the propellant weight and gross weight then decrease. For rockets having high payloads

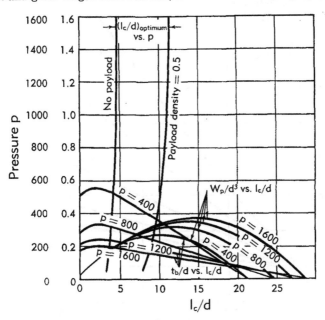

Fig. H,13f. Rocket motor characteristics for "fast burning" propellant with $r_{1000psi} = 0.75$ in./sec (see Fig. H,13b).

(for example, a rocket used for boosting a missile), a considerable decrease in l/d is possible without sacrificing much performance (see Fig. H,13c). In general, the smallest diameter rocket possible should be used—consistent with payload and burning time required—to obtain the most economical rocket.

H,14. Design of Propellant Grains.

Design procedure. After the specifications for thrust, burning time, and chamber pressure have been determined, the details of rocket motor design can be worked out. For the design of the propellant charge the basic propellant characteristics as described in Art. 12 must be known. Useful design curves which may be prepared from these basic characteristics are shown in Fig. H,14a, H,14b, and H,14c. These curves are prepared from data obtained from actual rocket motor tests. Theory, alone, does not give sufficiently precise practical design data.

The critical propellant characteristics which must be known are propellant weight, web thickness,[8] and burning surface. The propellant

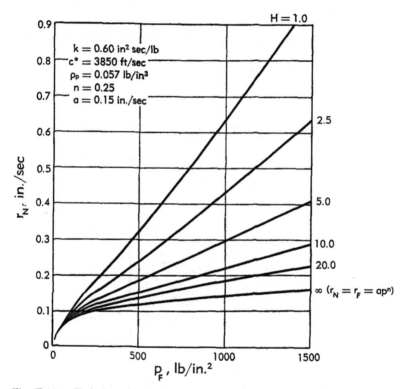

Fig. H,14a. Variation of burning rate, at the exhaust nozzle end of the charge, with front pressure and area ratio H, for a heterogeneous propellant (70°F). Note: $H = 1/J$.

weight can be calculated from the basic thrust equation $F = \dot{m}Vg$ by

$$W_p = \frac{gFt_b}{C_Fc^*}$$

where C_F may be estimated to be about 98 per cent of the theoretical value. The web thickness is determined simply by

$$w = rt_b$$

[8] The web thickness w is the thickness of propellant measured normal to the burning surface.

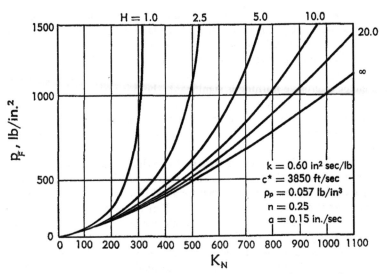

Fig. H,14b. Front-chamber pressure as a function of the area ratios K and H for a heterogeneous propellant (70°F). Note: $H = 1/J$.

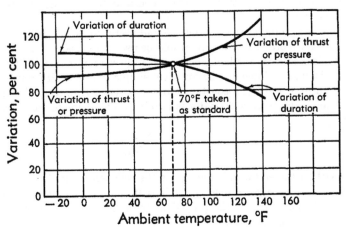

Fig. H,14c. Variation of performance with ambient temperature. (Erosive amplification is neglected.)

where the burning rate is selected from Fig. H,14a. Knowing the design pressure, the ratio $K = S_b/A_t$ can be found from Fig. H,14b; then the burning area is calculated by

$$S_b = KA_t = K\frac{F}{C_F p_o}$$

The variation of thrust and burning time vs. temperature is estimated from Fig. H,14c. Upon referring to Fig. H,14a and H,14b, the question of what gas duct-to-nozzle throat area ratio to use no doubt occurs to the viewer. For many propellants, erosive burning is negligible when $H > 5$, so that the curve for $H = \infty$ is often the only curve necessary. When the effect of erosive burning is significant, the pressure and burning rate variation with gas duct area change make it necessary to use curves such as Fig. H,14a and H,14b for a trial and error solution of w and S_b. Care must be exercised not to enter the region of erosive instability (see Art. 9).

Having the weight, burning area, and web thickness, the designer must determine the diameter and length dimensions which are dependent upon the particular design chosen. A discussion of charge geometry follows.

Propellant geometry. Because of the limited range of burning rates that exist for solid propellants, much work has been done with geometry to obtain design versatility. The propellant charge may burn on one end only (end-burning or "cigarette" burning), from the inside out (internal-burning), from the outside in (external-burning), or in any combination of the three (such as a completely unrestricted tubular charge). Charge shapes which have been found useful are shown in Fig. H,2a; all of these shapes can be designed to give nearly constant burning surface as the web burns away.

The end-burning charge has been found useful for long duration, low thrust rocket motors. It probably would be used universally if any burning rate desired were available, because it utilizes approximately 100 per cent of chamber volume and because it is simple to design: it exhibits no erosive burning and has a constant burning surface. However, with high thrust units, too large a diameter is required. The diameter required is given by

$$d_p = d_t \sqrt{K} = \frac{F\sqrt{K}}{C_F p_0}$$

The length of the end-burning charge is the web thickness. Another disadvantage of the end-burning charge is that the combustion gases flow past the chamber wall as the propellant burns away, and this requires heavy walls to absorb heat.

The double-base propellants were extruded into tubular shapes for use in artillery rockets during World War II. Because, for this design, the gases flow along the charge both internally and externally, it is necessary to so design that the pressure outside the charge is approximately equal to that inside.[9] This condition requires that the ratio of burning area

---

[9] In some designs it has been found necessary to drill radial holes through these charges to help equalize the inside and outside pressures, as well as to eliminate resonant burning effects.

to gas duct area be the same for the inside and outside channels, or expressed as a condition on the mean diameter z of the tubular propellant charge [57].

$$\frac{z}{d_c} = \frac{1}{2}\left[\left(\frac{w^2}{d_c^2} + 2\right)^{\frac{1}{2}} - \frac{w}{d_c}\right]$$

The corresponding limit on the length of the charge can be expressed by

$$\frac{l_p}{d_c} = \frac{12.5}{(z/d_c)} - \frac{50w}{d_c}$$

These limitations on charge diameter and length are plotted vs. web thickness in Fig. H,14d.

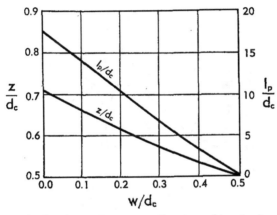

Fig. H,14d. Optimum mean diameter and length of
single tubular grains of a given web thickness.

The simplicity of design and manufacture are the principal advantages of tubular charges. Some of the disadvantages are:

1. The gas flow between the charge and chamber wall produces high heat input to the chamber. As a consequence, the chamber wall gets hot quickly, making long duration rockets impractical because of the heavy wall required.
2. High volumetric efficiency is impossible, values greater than 65 per cent usually being rare. Thus low impulse weight ratios are characteristic of all motors using tubular charges.
3. Complicated arrangements are often necessary to hold these charges in place in the rocket motor.
4. Impulse is often lost by charge breakup near the end of burning when the web is very thin.

⟨ 149 ⟩

The cruciform charge has been found slightly superior to the tubular charge with respect to volumetric loading and the ability to withstand deformation, but it has the disadvantage of burning externally. Both the cruciform and tubular charges have been discussed in detail in [10].

The internal-burning charges such as the star, rod-and-tube, slotted-cylinder, multiperforated, and multidisk have the very attractive feature that the propellant itself insulates the combustion chamber wall from the hot combustion gases. With available burning rates, high thrust, short duration units can be designed with the multiperforated or multidisk

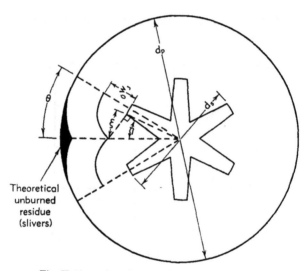

Fig. H,14e.   Notations employed in the design of
internal-burning, star-center propellant charges.

charges or long duration and low thrust can be achieved by using the star, slotted-cylinder, and rod-and-tube shapes. Also, high volumetric efficiencies (80 per cent and higher) can be achieved. As a consequence, the trend is to use internal-burning designs in most applications.

A geometric study of cross-sectional loading and the burning surface as a function of web thickness is necessary before preparing to design these charges. For illustration consider a study that has been made [57] of the internal star shape. If the star parameters are defined as given in Fig. H,14e, the variation in gas duct area and burning surface per unit length of charge (perimeter) vs. the amount of web burned may be expressed by

$$_0(A_d)_J = A_{d0} + \left(\frac{S_b}{d_p}\right)_0 w + \frac{M}{2} w^2$$

Note here that w refers to linear displacement of burning surface from its original position (web distance burned).

$$_J(A_d)_f = N\left[\left(\frac{d_s}{2} + w^2\right)(\theta - \xi) + w^2\left(\xi + \sin^{-1}\frac{d_s\sin\xi}{2w}\right)\right.$$
$$\left. + \frac{1}{2}d_s\sin\xi\left(R\cos\xi + w\cos\sin^{-1}\frac{d_s\sin\xi}{2w}\right)\right]$$

$$_0\left(\frac{S_b}{l_p}\right)_J = \left(\frac{S_b}{l_p}\right)_0 + Mw$$

$$_J\left(\frac{S_b}{l_p}\right)_f = 2N\left[\frac{1}{2}(\theta - \xi)d_s + w\left(\theta + \sin^{-1}\frac{d_s\sin\xi}{2w}\right)\right]$$

where

$$\left(\frac{S_b}{l_p}\right)_0 = Nd_s\left(\theta - \xi\frac{\sin\xi}{\sin\eta}\right) = \text{initial perimeter}$$

$$M = 2N\left(\frac{\pi}{2} + \theta - \eta - \cot\eta\right)$$

$$= \text{initial slope of perimeter versus web distance burned}$$

The subscript $_0$ refers to the initial condition and $_0(\ )_J$ indicates the interval over which the "flat" sides of the inner star points exist. After the "flat" sides disappear, the propellant burns outward until the motor wall is contracted; this interval is designated by $_J(\ )_f$. The junction point where the "flat" sides disappear occurs when the web distance is

$$w_J = \frac{d_s}{2}\frac{\sin\xi}{\cos\eta}$$

When the motor wall is contacted, the amount of propellant left is called the "theoretical unburned residue." It is desirable to keep the theoretical unburned residue to a minimum because the actual unburned residue is often roughly proportional to the theoretical amount.

For quick estimation of design, the long expressions for A_d and S_b are not convenient, but graphical methods are most useful. For a given number of star points N and initial-to-final perimeter ratio λ, the volumetric loading $1 - \alpha$, unburned residue σ, minimum perimeter ratio as the web burns away μ, and values for critical angles ξ and η can be shown as a function of web fraction $1 - \delta$; an example is given in Fig. H,14f [58]. The use of these geometrical design curves is demonstrated in the following paragraphs which show how the pressure-time curve is calculated.

Calculation of the pressure-time curve. The minimum chamber wall thickness is obtained for a given average pressure if the pressure-time curve is constant. One may therefore conclude that the most efficient rocket motor should operate at a constant pressure; however, for an internal-burning star charge, for example, constant pressure (or constant burning surface under nonerosive burning conditions) can be obtained only at a sacrifice of high volumetric loading and low unburned propellant residue. The impulse-weight ratio of a rocket motor can often be improved by adding propellant weight and allowing the pressure to

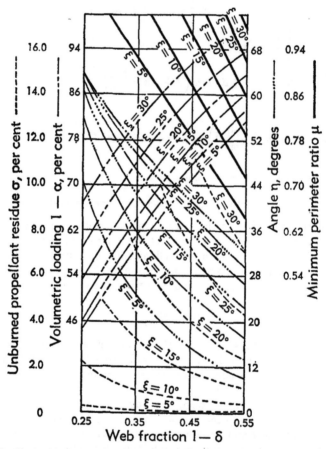

Fig. H,14f.   Relationship of geometrical variables for an internal-burning, star-center propellant charge (see Fig. H,14e). ($N = 6$, $\lambda = 0.80$.)

increase with time. Fortunately, the results for some applications, such as assisted take-off of aircraft and high altitude rockets, are improved with an increasing pressure-time curve. The port area is small when high propellant loading is obtained with internal-burning charges; as a consequence, the port-to-throat area ratio is often small, thus leading to high erosive burning.

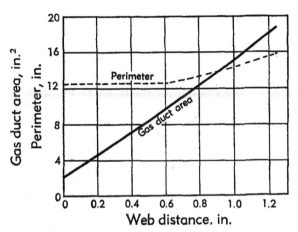

Fig. H,14g. Variation of perimeter and gas duct area with web distance for the star-center charge (see Fig. H,14e).

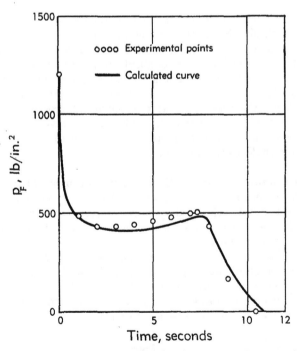

Fig. H,14h. Comparison between calculated and experimental pressure-time curve for an internal-burning, star charge (see Fig. H,14e and text).

An example of the calculation of a pressure-time curve for an internal-burning star charge will now be made. From the design chart of Fig. H,14f, a design is chosen which has a shape identical to the one of Fig. H,14e for this design $1 - \delta = 0.5$, $1 - \alpha = 89$ per cent, $\sigma = 5$ per cent, $\mu = 1$, $\lambda = 0.8$, $N = 6$, $\theta = 30°$, $\eta = 33.5°$, and $\xi = 24°$. The variation of port area and burning surface per unit length vs. web distance is plotted in Fig. H,14g. If the propellant chosen has the characteristics given by Fig. H,14a and H,14b, and then a charge diameter of 5 in., a length of 40 in., and a port-to-throat ratio of 2 are selected, all necessary information for the calculation is available. By the use of Eq. 9-1 the initial front chamber pressure and front burning rate are determined. The burning rate and pressure at the nozzle end of the charge are determined by Eq. 8-13, 8-15 and 9-1. Because the port area changes continually and the pressure and burning rate depend upon port-to-throat area ratio, incremental time intervals must be selected over which the calculated conditions are assumed to prevail. By successive recalculations at the end of each time interval, the pressure-time curve shown in Fig. H,14h is obtained. The pressure-time curve resulting from an actual test is also shown and the comparison between theory and experiment is evidently very good.

It is interesting to note that the shape of the pressure-time curve does not look much like that of the burning surface in Fig. H,14g; this non-similarity is due to the strongly erosive burning character of the propellant. It is also of interest that about 15 per cent of the impulse is obtained after the propellant burns through to the chamber wall at the nozzle end of the charge; thus the calculated propellant residue has little meaning when the propellant exhibits significant erosive burning. In order to achieve an abrupt cessation of burning, it may be necessary to taper the gas duct so that the propellant burns out to the chamber wall all along its length at the same time.

H,15. Mechanical Design. The mechanical components of a rocket motor are comprised of the combustion chamber, exhaust nozzle, and miscellaneous items such as safety release devices and charge supports. Views of the mechanical components of three typical motors are shown in Fig. H,15a, H,15b, and H,15c.

The combustion chamber. The combustion chamber is essentially a pressure vessel containing the propellant, which must be able to withstand the high pressure and temperature of the gases. The wall thickness τ of the tube which constitutes the main portion of the combustion chamber can be estimated by the hoop tension formula

$$\tau = \frac{p_{max}d_w}{2\sigma}$$

when the wall thickness computed is less than 5 per cent of the diameter, which is the usual case for well-designed motors. The chamber pressure upon which to design should be the maximum expected, plus a safety allowance. Normally, $p_{max} = p_o(1 + f_p)(1 + f_{bv})(1 + \pi_p\Delta T)f_s$, where f_p is a factor

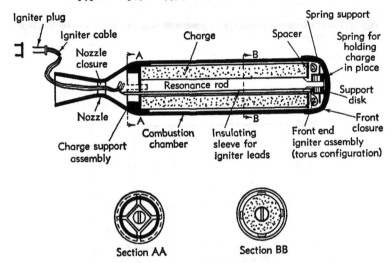

Fig. H,15a. Mechanical design characteristics of a large rocket employing a tubular charge. (After Jato Manual M1.)

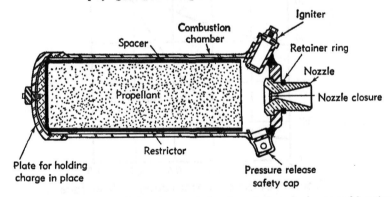

Fig. H,15b. Mechanical design characteristics of a rocket employing an end-burning charge. Note that propellant is not sealed to combustion chamber walls. (After Aerojet Jato design.)

indicating increase in chamber pressure above the average, f_{bv} is a batch variation factor for propellants, and f_s is a safety factor allowance. However, if the motor is spin-stabilized the effective increase in internal pressure caused by centrifugal force must also be considered; this may be estimated to be $\tau\omega^2 d_w\rho_w/2g$. The inner chamber diameter d_w is determined

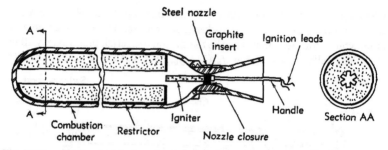

Fig. H,15c.   Mechanical design characteristics of a rocket motor employing a star-center, internal-burning charge sealed to combustion chamber wall.

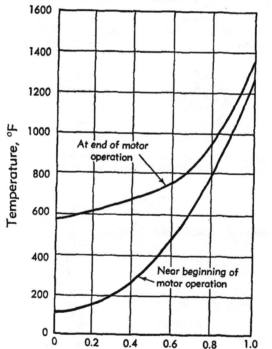

Fig. H,15d.   Illustration of temperature distribution in the combustion chamber wall of a rocket motor.

by the sum of the charge diameter, the restriction thickness, and the space required to allow for the thermal expansion and flow of gases (external-burning charges). The value of the tensile strength depends upon the effective temperature of the combustion chamber wall at the end of burning and the rate at which the motor wall is heated. Since a temperature distribution, such as that shown in Fig. H,15d, may be present in some rocket motor walls, it is obvious that a complicated stress distribution may exist due to both the chamber pressure and the thermal gradient. It is usually adequate, however, to use the mean wall temperature for estimating the strength of the material being used. It remains, then, to determine accurately the value of the mean wall temperature for any particular design study.

The transfer of heat from the hot gases to the combustion chamber can be expressed by

$$q = h(T_r - T_w)$$

where T_r is the recovery temperature of the gases in the motor, T_w is the temperature of the motor wall at the inner wall surface, and h is the heat transfer coefficient which may be expressed [10] as:

$$h = \frac{\text{const} \times c_p G^n}{D^{1-n}}$$

where the constant depends upon the viscosity and thermal conductivity of the gases, G is the mass flow density, and c_p is the specific heat of the gases at constant pressure, and D is the effective duct diameter through which the gases pass. For double-base and "perchlorate" propellants, the heat transfer coefficient along the cylindrical portion of the combustion chamber can be approximated [10] by[10]

$$h \cong 0.046 c_p G^{0.8}$$

Since the diameter variation is not usually large, $D^{0.2}$, it can be neglected without much error. The value of G can be determined easily at the nozzle end (where the wall temperature is highest) of a cylindrical combustion chamber from the pressure-time curve and charge design.

$$G = \frac{W}{A_d} = \frac{g}{c^* \theta^*} p_F J$$

As the chamber wall heats up, however, the value of T_w in the equation for q changes continually, thus causing the heat input to vary. A practical solution to this transient heat flow problem has been obtained by Hall [59] by applying the numerical method of Dusinberre to hollow cylinders. Analytic solution of the Poisson-Fourier equation, applicable to a cylindrical system having circular symmetry and no axial heat flow,

$$\frac{\partial T}{\partial t} = \kappa \left[\frac{\partial^2 T}{\partial R^2} + \frac{1}{R} \frac{\partial T}{\partial R} \right]$$

[10] The h's are greater than indicated by the equation when $G < 1200 p^{\frac{1}{2}}$, where p is the pressure in atmospheres. In the relationship shown here, h is expressed in BTU/ft²/hr/°F and G in lb/ft²/hr. In rocket work the use of in.² and sec may be more convenient than ft² and hr.

is discussed in [60]. The numerical method does not require mathematical skill, gives acceptable accuracy, and can easily take care of changing thermal properties; in contrast, most analytic solutions require laborious numerical integration. Tsien and Cheng have developed a useful similarity law for this problem [61]. If the chamber wall is very thin, of thickness τ_e, an approximate conservative solution can be obtained simply by considering the heat flow over very small time intervals, assuming the

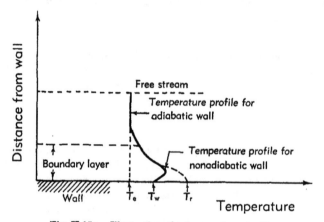

Fig. H,15e.   Illustration of a temperature profile
for gas flow along a solid boundary.

wall of heat capacity c_e to heat uniformally across its thickness, calculating a temperature rise, and then repeating the operation for the next time interval.with the new wall temperature. By equating the heat input to the heat absorbed,

$$q\Delta t = \rho_e \tau_e c_e \Delta T$$

or

$$\Delta T = \frac{q\Delta t}{\rho_e \tau_e c_e}$$

An example of how the wall temperature distribution curve may vary with time was given in Fig. H,15e.

The joints, attaching the front and rear closures to the chamber body, may provide a difficult design problem requiring detailed stress analysis. In general, the principal load on the joint is longitudinal. In addition to the pressure, the effects of acceleration, internal drag forces, and external drag forces must be considered.

The exhaust nozzle. The thrust of a rocket motor is significantly dependent upon nozzle design, as evident from the equation (see G,1)

$$F = C_F p_o A_t$$

where

$$C_F \cong 0.98 \left[\frac{1}{2} (1 - \cos \psi) \right] \left[\Gamma' \sqrt{\frac{2}{\gamma - 1}} \, \eta_{is} + \frac{p_e - p_\infty}{p_o} \frac{A_e}{A_t} \right]$$

The highest possible value of C_F is desired, consistent with low nozzle weight. The half angle of the exit section ψ should be between 8 and 20° for good nozzle performance. Very small angles make the downstream portion of the nozzle excessively long from the weight and heat transfer standpoint; in general $\gamma \cong 15°$ is satisfactory. The half angle of the inlet section of the nozzle β does not appear in the above equation. Although β is not critical from a thrust standpoint, a smooth contour for the inlet is desirable; in general $20° \leq \beta \leq 45°$ is satisfactory. The value of the

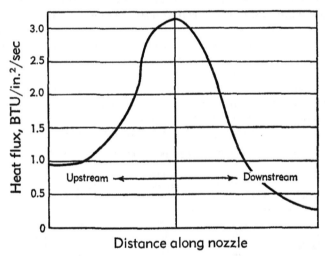

Fig. H,15f. The variation of heat flux with position along an exhaust nozzle.

expansion ratio A_e/A_t to give optimum thrust coefficient depends upon the particular application. If the exit pressure p_∞ is constant during the motor operation, the optimum value of A_e/A_t is [42]:

$$\frac{A_e}{A_t} = \frac{\left(\dfrac{2}{\gamma + 1} \right)^{\frac{\gamma+1}{2(\gamma-1)}}}{\left(\dfrac{p_\infty}{p_o} \right)^{\frac{1}{\gamma}} \sqrt{\dfrac{2}{\gamma - 1} \left[1 - \left(\dfrac{p_\infty}{p_o} \right)^{\frac{\gamma-1}{\gamma}} \right]}}$$

For $\gamma = 1.2$, $p_o = 2000$ lb/in.², and $p_\infty = 14.7$ lb/in.², the value of $A_e/A_t = 15$ for optimum C_F.

The heat transfer problem in the exhaust nozzle is severe, since the mass flow density G is greatest at the nozzle throat. An example of the heat flux vs. position along nozzle length is illustrated by Fig. H,15f.

As with the combustion chamber, the heat input to the nozzle can be given by the equation $q = h(T_r - T_w)$. However, the heat transfer coefficient, which applies for the tubular combustion chambers, does not apply to exhaust nozzles. The results of experimental tests at 500 psi [59] have indicated that the following relationships may apply to nozzles:[11]

Entrance portion of nozzle (subsonic flow): $h \cong 2.16 c_p G^{0.5}$

Exit section of nozzle (supersonic flow): $h \cong 0.00106 c_p G$

By the use of these equations, the temperature rise in a given nozzle can be estimated in a manner similar to that described for the combustion chamber.

An example of the effect of h upon the variation of temperature with time at the throat section of a copper nozzle is shown in Fig. H,15g[12]. The temperature of the metal at the surface next to the gas stream is of particular interest because, as it approaches the melting point, the nozzle erodes. In order to keep the inner wall temperature to a minimum, the thermal conductivity should be high. To keep the nozzle volume and weight small, the density and specific heat should be high.

Molybdenum, graphite, copper, and iron have been found to be the most useful materials for the heat-absorbing nozzle. Some of their properties are listed in Table H,15. Molybdenum and graphite have very high

Table H,15. Approximate properties of some nozzle materials.

Material	Thermal* conductivity, BTU/ft hr °F	Specific* weight, lb/ft³	Specific* heat, BTU/lb °F	Melting point, °F
Copper	223	560	0.09	1981
Graphite	80	97	0.16	6600 (sublimes)
SAE 1020 steel	30	491	0.11	2760
Iron	46	487	0.11	2795
Molybdenum	84	636	0.06	4748

* At 70°F.

melting (or subliming) temperatures and generally make the lightest nozzles. Unfortunately, molybdenum is not plentiful or easy to fabricate in large sizes. Graphite has very low strength and, consequently, must be used as an insert (Fig. H,15c). For combustion periods less than 5 sec, iron may be most practical. Copper is satisfactory for longer periods (30–60 sec), but its weight is usually prohibitive.

Heat-resistant materials offer the possibility of lighter weight than

[11] In these equations h is expressed as BTU/ft² hr °F and G as lb/ft² hr.
[12] Fig. H,15g suggests a convenient method for determining the value of h: by merely recording the temperature of the outside wall for a given section, which is insulated (such as by air space) against axial heat flow, h can be estimated by fitting the experimental temperature-time curve on a set of curves calculated for several values of h.

that obtained with heat-absorbing nozzles, but, in general, they are not of adequate strength to withstand thermal stresses. Silicon carbide has been used for some uncooled nozzles. A comprehensive treatment of exhaust nozzle materials is given in [62].

Charge support. When the charge is not sealed to the combustion chamber it usually must be held in place by some mechanical means.

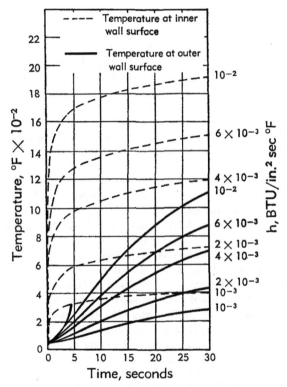

Fig. H,15g. Computed variation in temperature at the throat section of an exhaust nozzle, with time for several values of the heat transfer coefficient H. (Gas temperature 3200°F, copper nozzle, throat diameters: inside 1.28 in., outside 5.50 in.)

External-burning charges, in particular, sometimes require elaborate methods of support which may add considerable weight to the rocket motor. The supports must be designed to withstand the inertia forces and drag forces of gases such as those discussed in Art. 13, and also the high temperature combustion gases. Methods of supports for a few charge designs are illustrated in Fig. H,15a, H,15b, and H,15c.

Pressure release device. When a rocket motor is to be used near personnel, it usually has a safety device to release the chamber pressure if it should approach the maximum allowable design value. Copper diaphragms have been found useful as burst diaphragms, because copper can be obtained easily in high purity. When used, copper must be insulated from the combustion gases. For a rupture disk of high purity (99.5 per cent) copper, the burst pressure is given by [42]

$$\text{burst pressure, lb/in.}^2 = \frac{98,400(t - 0.0035)}{d}$$

where t = disk thickness, inches, and d = shear area diameter of disk, inches. This empirical formula is satisfactory over the pressure range 1500–4000 lb/in.² for disks from 0.6 to 2.0 in. in diameter.

In general, the gases which escape through the pressure release device should be directed so that they produce no net thrust on the rocket motor.

Ignitors. The purpose of the ignitor is to bring the exposed propellant surface to the ignition temperature as quickly as possible without excess pressure peaks, which might rupture the motor or charge. The following remarks refer to practical methods, although the theory of ignition [63] is developing rapidly.

The principal components of the typical ignitor consist of a squib, ignition granules, and the ignitor case. The squib usually consists of an electrical wire heater imbedded in a small amount of heat-sensitive primary ignition. When ignited, the squib flashes to ignite the main ignition charge. The gases from the main ignition charge burst the container and throw the ignited particles against the propellant surface. For minimum delay time, the ignition particles should be at high temperature and should cover as large a portion of propellant surface as possible. A typical ignition history for a rocket motor is shown in Fig. H,15h.

The principal factors which determine the ignition characteristics are the type, granulation, amount, moisture content, and packing of the ignition powder; the location and construction of the ignitor case; the charge geometry; the ignition temperature of the propellant; the surface condition of the propellant (the presence of inert films such as wax from mold release); and the relationship between ignition pressure peak, nozzle closure blowout pressure, and the operating pressure for the motor. Since many of these variables are interdependent, ignitor design is at the mercy of the experience and judgment of the designer. Some qualitative guides for ignitor design may be obtained from [10] and the remaining portion of this section.

Not much attention has been given to the preparation of new ignition powder. The most commonly used material is black powder, which is readily available in several granulations. For some special applications, mixtures of metals with oxidizing agents (such as magnesium and aluminum with potassium perchlorate) have been useful; the combustion of such mixtures produces high temperatures, relatively small amounts of gas, and ignition times less than those obtained from black powder. The metal perchlorate mixtures are often extremely hazardous, and are used only for special experimental studies.

The ignition time can usually be kept to a minimum by using a nozzle closure to prevent the gases from exhausting during the ignition period. An approximate rule is to have the pressure from ignition be about 30 to 40 per cent of the expected motor pressure, and the nozzle closure blow-out pressure be about 90 to 100 per cent of the expected motor pressure. For black powder, the number of grams of ignition (W_i) required to raise

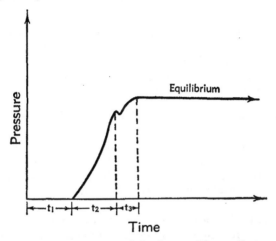

Fig. H,15h. Typical ignition characteristics for a solid propellant rocket motor. t_1 = interval between application of electric current and pressure rise = 20 milliseconds, t_2 = interval between rise of pressure and nozzle closure blowout = 50 milliseconds, t_3 = interval between nozzle closure blowout and establishment of equilibrium = 20 milliseconds.

the pressure (lb/in.^{2}) to a given value in a given volume v (in.^{3}) may be found from

$$p = 2000 \left(\frac{W_i}{v}\right)^{0.86}$$

If the amount of ignition is too high, the nozzle closure blows out before enough of the propellant can burn to sustain combustion, resulting in a misfire.

The ignitor is usually located at the nozzle end of the motor, the head end, or distributed along the length of the charge; two of these types are illustrated in Fig. H,15a, H,15b, and H,15c. The ignitor which is located in the nozzle of the motor is usually economical, because it can be made simply and because no special machining of the combustion chamber is necessary to hold it. One objection to this type is that the flying fragments from the ignitor may be destructive. For internal-burning charges, a "shotgun" blast from the nozzle-end ignitor into the central perforation provides effective ignition. For external-burning charges, the forward-end ignitor has been popular because the ignitor flame must pass over the exposed propellant surface to escape through the nozzle. The ignitor which is distributed along the length of the charge is desirable when no nozzle closure is allowable; a "flash" type of ignition (such as the metal-perchlorate mixture) is often useful in this type of ignitor.

CHAPTER 5. DEVELOPMENT TRENDS

H,16. Trends in Solid Propellant Rocket Development.

Practical trends. Present efforts in rocket development are directed toward achieving flexibility in design, high performance, reliability, and low cost. In summary, the major factors which are furthering the attainment of these ideal goals are as follows:

1. Internal-burning charge design. (Combustion chamber case insulated from hot gases.)
2. Propellants with burning rates insensitive to pressure charges. (Operating conditions insensitive to temperature change and design factors.)
3. Casting processes. (Large and small charges made with nearly equal ease. Cost possibly much lower than high pressure molding and extrusion.)
4. High energy, high density propellants. (Energy per unit volume and per unit weight high, results in decreased over-all motor weight and, usually, decreased cost.)
5. Motor construction materials of high strength and low weight. Improved fabrication techniques.
6. High strength, low conductivity, heat-resistant materials. (Allows low weight for parts exposed to combustion gases, such as the exhaust nozzle.)
7. Improved quality control to enhance reliability and useful applications.

Design flexibility depends a great deal upon the ability to obtain a wide range of burning rates and still retain smokeless, noncorrosive exhaust gases; however, much work must still be done in order to combine these desirable characteristics. The trend at present is to use complex charge geometry, when high mass flows are required, to compensate for the low burning rate propellants which are available.

The cost of solid propellants has been high, at the very best, due to the safety requirements necessary in their manufacturing processes and the high basic material cost. The trend is to attempt to use less expensive materials in propellants for some applications, even though relatively low performance results. Because of the infancy of some processes—the cast process, a notable example—streamlining of manufacture has not been perfected. Regardless of the high propellant charge cost ($1 to $5 per pound being a reasonable range in 1948), the motor construction is so simple that the over-all cost of the complete solid propellant rocket unit may be less than a comparable liquid propellant rocket, which uses much less expensive propellant materials. Furthermore the simplicity of operation makes the solid propellant rocket motor particularly desirable for field use. In addition to rocket propulsion, the gas generation capabilities of solid propellants indicate that they may be used for many novel applications in the future.

Fundamental problems. The fundamental problem of solid propellant rockets is the detailed, quantitative understanding of the combustion process under a wide variety of conditions. Ultimately, one may hope to control the burning rate over a wide range of speeds by means of propellant composition and processing.

Understanding of the combustion process will require clarification of many diverse scientific problems, such as chemical kinetics, diffusion of heat and molecular species, radiation, the behavior of small particles in a gas stream and at a solid surface, and the hydrodynamic problem of gas flow past a burning surface.

A second class of fundamental problems concerns the control of propellant quality during manufacture. Relatively small changes may have large effects upon rocket motor operation. The technology of mixing small particle, high polymers and the vast arts of chemical engineering will require continued application and development.

H,17. Cited References.

1. Hayes, T. J. *Elements of Ordnance.* Wiley, 1938.
2. Symposium on kinetics of propellants. *J. Phys. and Colloid Chem.* 54, 847 (1950).
3. Boys, S. F., and Corner, J. *Proc. Roy. Soc. London* A197, 90 (1949).
4. Corner, J. *Proc. Roy. Soc. London* A198, 388 (1949).
5. Corner, J. *Theory of Interior Ballistics of Guns.* Wiley, 1950.
6. Corner, J. The effect of turbulence on heterogeneous reaction rates. *Trans. Faraday Soc.* 43, 635 (1947).
7. Birkoff, G., MacDougall, D. P., Pugh, E. M., and Taylor, G. *J. Appl. Phys.* 19, 363 (1948).
8. Davis, L., Jr., Blitzer, L., and Follin, J. W., Jr. *Exterior Ballistics of Rockets.* McGraw-Hill. To be published.
9. Rosser, J. B., Newton, R. R., and Gross, G. L. *Mathematical Theory of Rocket Flight.* McGraw-Hill, 1947.
10. Wimpress, R. N. *Internal Ballistics of Solid-Fuel Rockets.* McGraw-Hill, 1950.
11. Grad, H. *Commun. on Pure and Appl. Math.* 2, 79 (1949).
12. *Fourth Symposium on Combustion.* Williams & Wilkins, 1953.
13. *Fifth Symposium on Combustion.* Reinhold, 1955.
14. Cheng, S. I. High-frequency combustion instability in solid propellant rockets. *J. Am. Rocket Soc.* 24, 27–32, 102–109 (1954).
15. Cheng, S. I. An unstable burning of solid propellants. *J. Am. Rocket Soc.* 25, 79 (1955).
16. Green, L., Jr. Unstable burning of solid propellants. *J. Am. Rocket Soc.* 24, 252 (1954).

17. Crawford, B. L., Jr. Rocket fundamentals. *George Washington Univ. N. O. R. C. Div. 3, Sec. H, Office Sci. Research and Develop. Rept. 3711,* June 1944.
18. Price, E. W. Steady-state one-dimensional flow in rocket motors. *J. Appl. Phys. 23,* 1942 (1952).
19. Price, E. W. Theory of steady state flow with mass addition applied to solid propellant rocket motors. *J. Am. Rocket Soc. 23,* 237 (1953).
20. Price, E. W. Charge geometry and ballistic parameters for solid propellant rocket motors. *J. Am. Rocket Soc. 24,* 16 (1954).
21. Price, E. W. One-dimensional, steady flow with mass addition and the effect of combustion chamber flow on rocket thrust. *J. Am. Rocket Soc. 25,* 61 (1955).
22. Sutton, G. P. *Rocket Propulsion Elements.* Wiley, 1949.
23. Seifert, H. S., Mills, M. M., and Summerfield, M. *Am. J. Phys. 15,* 1 (1947).
24. Avery, W. H. Radiation effects in propellant burning. *J. Phys. and Colloid Chem. 54,* 917 (1950).
25. Beek, J., Jr., Avery, W. H., Dresher, M. J., McClure, F. T., and Penner, S. S. Studies of radiation phenomena in rockets. *Office Sci. Research and Develop. Rept. 5817,* 1946.
26. Penner, S. S. *J. Appl. Phys. 19,* 278, 392, 511 (1948).
27. Avery, W. H., and Hunt, R. E. Effect of pressure and temperature on the rate of burning of double-base powders of different compositions. *Office Sci. Research and Develop. Rept. 1993,* Oct. 1943.
28. Crawford, B. L., Jr., Huggett, C., and McBrady, J. J. Observations on the burning of double-base powders. *Office Sci. Research and Develop. Rept. 3544,* Apr. 1944.
29. Gibson, R. E. Introductory remarks, Symposium on kinetics of propellants. *J. Phys. and Colloid Chem. 54,* 847 (1950).
30. Crawford, B. L., Jr., Huggett, C., and McBrady, J. J. The mechanism of burning of double base propellants. *J. Phys. and Colloid Chem. 54,* 854 (1950).
31. Crawford, B. L., Jr., Huggett, C., Daniels, F., and Wilfong, R. E. Direct measurement of burning rates by an electric timing method. *Anal. Chem. 19,* 630 (1947).
32. Thompson, R. J., and McClure, F. T. Erosive burning of double base powders. *Office Sci. Research and Develop. Rept. 5831,* Dec. 1945.
33. Wilfong, R. E., Penner, S. S., and Daniels, F. An hypothesis for propellant burning. *J. Phys. and Colloid Chem. 54,* 863 (1950).
34. Klein, R., Mentser, M., von Elbe, G., and Lewis, B. Determination of the thermal structure of a combustion wave by fine thermocouples. *J. Phys. and Colloid Chem. 54,* 877 (1950).
35. Rice, O. K., and Ginell, R. The theory of burning of double-base rocket powders. *J. Phys. and Colloid Chem. 54,* 885 (1950).
36. Parr, R. G., and Crawford, B. L., Jr. A physical theory of burning of double-base rocket propellants. Part 1. *J. Phys. and Colloid Chem. 54,* 929 (1950).
37. Boys, S. F., and Corner, J. Unpublished work.
38. Green, L., Jr. Erosive burning of some solid composite propellants. *J. Am. Rocket Soc. 24,* 9 (1954).
39. Muracour, H. Sur les lois de combustion des poudres colloidales. *Bull. soc. chim. France 41,* 19–51 (1927).
40. Touchard, M. L. Sur les anomalies de vivacité des poudres tubulaires. *Mém. artillerie franç. 26,* 297 (1952).
41. Longwell, P. A., Sage, B. H., and Lacy, W. N. The temperature of spontaneous ignition of several samples of American ballistite. *Natl. Defense Research Comm., Div. A, Sec. H, Calif. Inst. Technol. JDC 8,* Apr. 1942.
42. *Jet Propulsion.* Prepared by Staffs of Jet Propul. Lab. and Guggenheim Lab., Calif. Inst. Technol., for Air Tech. Service Command, 1946.
43. Shapiro, A. H., and Hawthorne, W. R. The mechanics and thermodynamics of steady one-dimensional gas flow. *J. Appl. Mech. 14,* A317–A336 (1947).
44. Geckler, R. D., and Sprenger, D. F. The correlation of interior ballistic data for solid propellants. *J. Am. Rocket Soc. 24,* 22 (1954).

45. von Kármán, Th., and Malina, F. J.   Characteristics of the ideal solid propellant rocket motor. *Calif. Inst. Technol. Jet Propul. Lab. Rept. 1–4,* 1940.
46. Mills, M. M.   The preparation and some properties of an asphalt base solid propellant Galcit 61-C. *Calif. Inst. Technol. Galcit Project 1, Rept. 22,* May 1944.
47. Vand, V.   *J. Phys. and Colloid Chem. 52,* 277, 300 (1948).
48. Dalla Valle, J. M.   *Micromeritics.* Pitman, 1943.
49. Furnas, C. C.   *Ind. Eng. Chem. 23,* 1502 (1931).
50. Mooney, M.   *J. Colloid Sci. 6,* 162 (1951).
51. Steinour, H. H.   *Ind. Eng. Chem. 36,* 618, 840, 901 (1944).
52. Hawksley, P. G. W.   *Nature 164,* 220 (1949).
53. Processing of rocket propellants. *Calif. Inst. Technol. Office Sci. Research and Develop. Rept. 2552,* 1946.
54. McClure, F. T., Rosser, J. B., and Kincaid, J. F.   The drag of the propellant gases on the power charge in rockets. *Office Sci. Research and Develop. Rept. 5872,* Feb. 1946.
55. Altman, D.   The thermal diffusivities of some solid propellants. *Calif. Inst Technol. Propul. Lab. Progress Rept. 9-32,* Apr. 1949.
56. Thackwell, H. L.   Unpublished data. *Calif. Inst. Technol. Jet Propul. Lab.*
57. Avery, W. H., and Beek, J., Jr.   Propellant charge design of solid fuel rockets. *Office Sci. Research and Develop. Rept. 5890,* June 1946.
58. Shafer, J. I.   Unpublished data. *Calif. Inst. Technol. Jet Propul. Lab.*
59. Hall, K. P.   Unpublished data. *Calif. Inst. Technol. Jet Propul. Lab.*
60. Perry, R. L., and Berggren, W. P.   Transient heat conduction in hollow cylinders after sudden change of inner-surface temperatures. *Univ. Calif. Publs. in Eng.,* 1944.
61. Tsien, H. S., and Cheng, C. M.   A similarity law for stressing rapidly heated thin-walled cylinders. *J. Am. Rocket Soc. 22,* 144 (1952).
62. Mills, M. M.   A study of materials for jet motor exhaust nozzles. *Air Corp. Calif. Inst. Technol. Jet Propul. Research, Galcit. Project 1, Rept. 18,* Aug. 1943.
63. Frazer, J. H., and Hicks, B. L.   Thermal theory of ignition of solid propellants. *J. Phys. and Colloid Chem. 54,* 872 (1950).

Milton Keynes UK
Ingram Content Group UK Ltd.
UKHW021043280924
448959UK00006B/138

9 780691 626185